Problem Solving in Physical Chemistry

∗

Problem Solving in Physical Chemistry

ROLAND R. ROSKOS

University of Wisconsin at La Crosse

WEST PUBLISHING CO.

St. Paul · New York · Boston · Los Angeles · San Francisco

COPYRIGHT © 1975 By WEST PUBLISHING CO.
All rights reserved
Printed in the United States of America
Library of Congress Catalog Card Number: 74–24649
ISBN: 0–8299–0028–4

Preface

    This book has been written to introduce the physical chemistry student to the fundamentals of computer programming so that he can use "canned" programs and so that he can write his own programs to complete more meaningful problems in physical chemistry. The book teaches the fundamentals of computer programming in the BASIC language, and it describes, in some detail, the running of the BASIC computer programs on time-shared computers that use interactive terminals. It assumes that students have no previous experience in writing or running computer programs. The book is written so that a college sophomore or junior can learn to use the computer by reading this supplementary text and by working problems related to his physical chemistry assignments. It is intended that little or no lecture time need be sacrificed in implementing the time-shared BASIC computer in physical chemistry classes.

    The first three short chapters are concerned with the writing of simple computer programs, with the running of these programs on the computer, and with the debugging and editing of these programs in the pursuit of satisfactory computer output. The sample programs within the chapters, and the programming assignments all deal with concepts and calculations which are likely to be found in the first week of a physical chemistry class. Step-by-step instructions are given for the first computer RUNS. Detailed descriptions are given for using computer commands, for using program storage devices, and for organizing and debugging computer programs.

The second section presents five canned programs which find frequent use in all of physical chemistry. These programs are intended permanent residents of a computer program library. Each of five chapters describes a single program, it lists the program, and presents step-wise instructions for the use of the program. The use of each of the canned programs is illustrated with a real physical chemistry problem. Exercises are also taken directly from physical chemistry topics. The PLOT program plots any continuous function via teletypewriter; AREA carries out numerical integration by use of Simpson's rule; LINEQ is a linear least squares fit of x-y data points; POLEQ is a polynomial least squares fit of x-y data points; and ROOTS finds all of the roots of most polynomials.

The third major section in this book contains approximately 200 different computer oriented problems divided into six chapters according to major topics within physical chemistry: Gas Calculations, Molecular Energies, Classical Thermodynamics, Statistical Thermodynamics, Kinetics, and Quantum Mechanics. Each chapter is divided into approximately ten subtopics. The concepts and equations from each subtopic are summarized; several computer problems follow each subtopic. Some computer problems require the use of previously described, frequently-used, canned programs; some require the use of other canned programs listed in the problem chapters; and many problems require simple student-prepared programs. An average student can easily complete a majority of the computer problems by spending approximately three hours per week with a computer terminal. Each of the problem chapters has been developed to be independent of the other chapters. The problem chapters may be used or covered in any order.

The final section in the book contains selected answers to the problems and exercises. These should prove to be particularly useful to the student as test cases for his newly written computer programs. If student answers coincide with the published answers, then he can be fairly confident that his program is working properly.

This textbook is intended to be a first exposure to computers and to computer programming. It is intended to be a source of physical chemistry problems which will supplement and in some cases replace problems which are found in physical chemistry texts. It is hoped that chemistry departments will develop additional canned programs and additional computer problems to meet the needs of individual courses in physical chemistry lecture and laboratory. It is the hope of this author that this textbook helps to integrate the use of computers with the learning and application of physical chemistry.

The author wishes to thank Professor Thomas Boates (Bemidji State College) and Professor Dale Nimrod (Luther College) who read the manuscript and who made many useful and constructive suggestions. Special thanks are in order to Mr. Leighton Lewis who wrote and tested most of the "canned" programs. The author also thanks Mrs. Judy Lathrop who patiently and diligently typed several versions of the manuscript.

Roland R. Roskos

*To
Beth, Michael
and Nancy*

Contents

1. **SIMPLE BASIC PROGRAMMING** *1*

    Introduction 1. Statements and Programs 2. Variables 3. Operators 3. E-Notation 4. An Improved Program 4. GO TO Statements 5. IF-THEN Statements 5. A More Sophisticated Program 6. READ, DATA Statements 6. FOR-NEXT Statements 7. The Use of Parentheses 8. Exercises 9.

2. **USING THE COMPUTER** *11*

    Introduction 11. Teletype Keyboard 12. Selected Commands for Computer 13. Running Sample Programs 15. Storing Programs for Future Use 18. Exercises 20.

3. **DEBUGGING** *21*

    Introduction 21. Program Organization 21. An Organized Example 23. Test Cases 25. Tracing 26. Error Statements 27.

4. **THE PROGRAM PLOT** *30*

    Introduction 30. Standard Functions 30. Defined Functions 32. Using the PLOT Program 32. Sample Calculation 36. Exercises 38.

5. **THE PROGRAM AREA** *39*

   Descriptive Comments Within Programs 39. Description of AREA 41. Using the Program AREA 43. Sample Calculation 45. Difficulties in Using AREA 46. Exercises 49.

6. **THE PROGRAM LINEQ** *50*

   Description of LINEQ 50. Numeric Data in Arrays 53. Subprograms 55. Using LINEQ 57. Sample Calculations with LINEQ 61. Exercises 63.

7. **THE PROGRAM POLEQ** *65*

   Description of POLEQ 65. Using POLEQ 66. Sample RUN with POLEQ 70. Exercises 72.

8. **THE PROGRAM ROOTS** *74*

   Description of ROOTS 74. Using ROOTS 76. Sample Calculation 80. Exercises 82.

9. **GAS CALCULATIONS** *83*

   Gas Densities from Ideal Gas Law 83. Van der Waals Equation of State 84. Virial Equation for Real Gases 85. Beattie-Bridgeman Form of Virial Equation 87. Critical State 89. Curve Fitting on Experimental Data 91. Boltzmann Distribution Function 92. Collision Properties of Gases 93.

10. **MOLECULAR ENERGIES** *95*

    Molecular Vibrations and Rotations 95. Boltzmann Distribution Law—Rotations of Molecules 101. Boltzmann Distribution Law—Vibrations of Molecules 103. First Einstein Function 105. Molecular Energies of Polyatomic Molecules 106. Heat Capacities of Ideal Gases 107.

11. **CLASSICAL THERMODYNAMICS** *109*

    Experimental Heat Capacities 109. Heat Capacities— Molecular Model 111. Gas Expansions and Compressions 112. Heating Gases 113. Enthalpies of Gases 113. Third Law Entropy 116. Entropies for Reactions 118. Free Energies of Reactions 119. Free Energy and Pressure 120. Equilibrium Constants 124. Equilibrium Concentrations 125. Temperature Dependence of Kp 126.

12. **STATISTICAL THERMODYNAMICS** *128*

    Translational Partition Function 128. Rotational Partition Function 128. Rotational Partition Function 131. Vibrational Partition Function 133. Molar Entropies of Ideal Gases 136. Free Energies and Equilibrium Constants 138. Perfect Crystal Model 139.

13. **KINETICS** *142*

First Order Reactions 142. Second Order in One Reactant 145. Second Order—Two Reactants 147. Pseudo Order Rate Constants 149. Differential Method for Reaction Orders 151. Activation Energies 153. An Interactive Kinetics Program 154.

14. **QUANTUM MECHANICS** *160*

The Particle in a Box 160. The Simple Harmonic Oscillator 163. The Hydrogen Atom 167.

SELECTED ANSWERS TO EXERCISES *172*

INDEX *185*

CHAPTER 1

Simple Basic Programming

1.1 INTRODUCTION

A simple problem from the first chapter of nearly any physical chemistry text may serve as an excellent example with which to demonstrate some simple programming in the Basic Language. The compressibility factor, Z, for a gas may be described by the following polynomial:

$$Z = 1 + AP + BP^2 + CP^3$$

If the gas under consideration is ethane and pressures are expressed in units of atmospheres, then the coefficients at $300°K$ are as follows:

$$A = -7.23 \times 10^{-3}$$
$$B = 15.0 \times 10^{-6}$$
$$C = 4.02 \times 10^{-9}$$

When investigating the compressibility of ethane over a pressure range, 0-500 atm, a chemist may wish to evaluate the above polynomial at 10 atmosphere increments over the entire pressure range. This involves 50 separate repetitive calculations which may be well suited for a computer calculation. The program which follows accomplishes the necessary calculations.

```
10 LET A = -.00723
20 LET B = .000015
30 LET C = 4.02E-09
40 INPUT P
50 LET Z = 1 + A*P + B*P↑2 + C*P↑3
60 PRINT Z
70 END
```

The above program then may be entered into the computer via teletype. When an appropriate command is given, the computer will execute the program or complete the calculations as designated by statements in the program. Statement 10 assigns the value of A in the polynomial; Statement 20 assigns the value of B in the polynomial and Statement 30 assigns the value of C in the polynomial. When the computer reaches Statement 40, it must ask the program operator for a value of pressure. After the operator submits a value of P via teletype, the computer will execute Statement 50, it will calculate the value of the compressibility at the just assigned pressure and it will store the numerical answer under the variable name Z. Statement 60 instructs the computer to print out, via teletype, the value of Z. Statement 70 informs the computer that the program has been completed. Notice that this program would need to be RUN 50 different times to obtain values for the compressibilities of ethane at pressures of 10 atm, 20 atm, 30 atm,···, 500 atmospheres, and that each time the program was executed, the program operator would be required to submit, via teletype, a new value of P.

1.2 STATEMENTS AND PROGRAMS

The program operator may communicate with the computer in a number of languages such as FORTRAN, COBOL, and BASIC. This book shall develop BASIC as an easy language by which students in Physical Chemistry may communicate with the computer. A program operator relates to the computer through STATEMENTS. A series of logical and consecutively numbered STATEMENTS produces a computer PROGRAM. The PROGRAM just described is a composite of 7 different STATEMENTS compiled in a logical sequence to define the compressibility of ethane at a defined pressure.

Each STATEMENT must begin with a STATEMENT number. Sequential numbers from 1-9999 may be used. Incremental STATEMENT numbers permit desired insertions when programs are edited or enlarged. The maximal length of a STATEMENT is 72 characters, the length of one teletypewriter line. A STATEMENT is frequently called a line.

1.3 VARIABLES

A variable is a symbolic name given to a quantity with a numerical value which can change or be calculated. Common variables in mathematics are given symbols such as X, Y, Z, T, S, and V. The variables in the polynomial for ethane are Z, A, B, C, P. These variables are then logically assigned the symbols Z, A, B, C, and P in the computer program. Quantities A, B, and C are considered fixed variables when completing calculations for ethane at $300°K$. The quantities P and Z are changing variables and assume a different value for each of the anticipated 50 iterations.

Any single letter of the alphabet can be called a variable, (A, B, C, D, $\cdots$ X, Y, Z). Any single letter followed by a single digit can also represent a variable (A1, A2, A3, A4, A5, A6, A7, A9, A0, B1, B2, ... Z7, Z8, Z9, Z0). These symbols allow 286 different variables in a single computer program. Some computer languages reserve the symbols with I, J, K, L for variables which are integers, numbers with no fractional component.

1.4 OPERATORS

OPERATORS are symbols incorporated into statements which direct the computer to perform an assigned task. The most fundamental operator is the mathematical analog of the equal sign. The symbol, = , may designate equality but more frequently designates that a variable is being assigned a numerical value. Proper operators (symbols) instruct the computer to add, subtract, multiply, divide, and exponentiate. Some useful OPERATORS are summarized below:

```
    =   :Value assignment to variable
    +   :Addition
    -   :Subtraction
    *   :Multiplication
    /   :Division
    ↑   :Exponentiation
    **  :Exponentiation (on certain systems)
```

The assignment operator, "=", assigns a fixed numerical value for A and B in STATEMENTS 10 and 20 of the above program. The program assigns Z a value as a result of several new operations. Careful inspection indicates that STATEMENT 50 is a mathematical evaluation of the polynomial.

Some simple examples illustrate further the use of operators.

$S = a + b$	10 LET S = A + B
$D = b - a$	10 LET D = B - A
$A = ax^2$	10 LET Y = A * X↑2
$V = \pi r^2$	10 LET V = 3.14 * R↑2
$V = abc$	10 LET V = A * B * C
$Y = mx + b$	10 LET Y = M * X + B
$Z = PV/RT$	10 LET Z = P * V / R / T

1.5 E–NOTATION

The computer accepts and prints out numbers with six significant digits. Numbers which are larger than .000001 and smaller than 100000. can be written in decimal form. Numbers outside the above limits must be expressed in exponential notation based on powers of 10. Therefore a number like 6,237,000,000 can be expressed as 6.237×10^9 which might be expressed as 6.237 * 10 ↑ 9 on the computer. To simplify the expression of large and small numbers the E notation has been developed. A typed E following a number signifies an exponential multiplier to follow. The numbers following the E indicate the power of ten. Therefore the above number would best be represented by 6.237 E 9. Finally the computer is capable of handling numbers between an approximate minimum of 10^{-38} to an approximate maximum of 10^{38}.

The examples which follow serve to illustrate the E format for expressing numbers:

6.023×10^{23}	6.023 E 23
6.62×10^{-33}	6.62 E-33
3×10^{8}	3 E 08
$.91 \times 10^{-27}$	.91 E-27
7.85×10^{-40}	not acceptable to computer
$8.54 \times 10^{+40}$	not acceptable to computer

1.6 AN IMPROVED PROGRAM

The program considered in Section 1.1 may be improved so that the compressibilities of ethane at 50 different pressures are calculated more efficiently. The program which follows illustrates additional features of the BASIC language.

Simple Basic Programming

```
10 LET A = -.00723
20 LET B = .000015
30 LET C = 4.02E-09
40 LET P = 0
50 LET Z = 1 + A*P + B*P↑2 + C*P↑3
60 PRINT P,Z
70 IF P = 500 THEN 100
80 LET P = P + 10
90 GO TO 50
100 END
```

This program is designed to calculate a large number of compressibilities in a single RUN. The variables A, B, and C are fixed in the first three STATEMENTS. STATEMENT 40 initializes the value of the variable P which is also the value of P used in the first calculation of compressibility. STATEMENT 90 returns the computer to STATEMENT 50 so as to calculate another compressibility at another pressure. Notice that before STATEMENT 90 is executed, STATEMENT 80 changes the value of P. The new value of variable P is the old value of P incremented by 10 atmospheres. Notice further that the computer places itself into an infinite cycle in that in each iteration a value of Z is calculated, the values of P and Z are printed, and the value P is incremented by ten units. Therefore a STATEMENT such as STATEMENT 70 is necessary. During each cycle, the computer inspects the value of P. If the value of P is not 500, then the program is allowed to continue in its cycles. When the value of P reaches 500, then the computer is directed to proceed to the next portion of the program beyond the GO TO STATEMENT. Thus the program completes the desired fifty calculations.

1.7 GO TO STATEMENTS

The GO TO STATEMENT allows the computer programmer to alter sequential execution of program STATEMENTS. Occasionally the GO TO STATEMENT allows the program to delete execution of a series of STATEMENTS. More frequently, however, the GO TO allows the programmer to repeat the execution of a series of program STATEMENTS.

The GO TO STATEMENT must be followed by an expression, usually a STATEMENT number. GO TO STATEMENTS should not be used to enter FOR-NEXT LOOPS which will be described later.

1.8 IF—THEN STATEMENTS

The computer has the potential to make logical decisions with the use of IF-THEN STATEMENTS. In the program from Section 1.6 the computer is asked to decide when to terminate the iterative calculation. On each cycle, the value of P is compared to 500. If the condition is not met (P does not equal 500) then the program is allowed to continue. But when the condition is fulfilled, then the execution is transferred to the line indicated by the IF-THEN STATEMENT.

An IF-THEN STATEMENT consists of a statement number, the word IF, the test condition, the word THEN, and a STATEMENT number which is used when the condition is fulfilled. The examples that follow illustrate other useful conditions for IF-THEN STATEMENTS:

```
70 IF  P = 500 THEN 100
70 IF  P > 500 THEN 100
70 IF  P > = 500 THEN 100
70 IF  500-P < 0 THEN 100
70 IF  500-P < = 0 THEN 100
```

1.9 A MORE SOPHISTICATED PROGRAM

```
10 READ A,B,C
20 FOR P = 0 to 500 STEP 10
30 LET Z = 1 + A*P + B*P↑2 + C*P↑3
40 PRINT P,Z
50 NEXT P
60 DATA -.00723, .000015, 4.02E-09
70 END
```

This program again evaluates the compressibilities of ethane at various pressures ranging from 0 to 500 atmospheres. STATEMENT 10 directs the computer to read the values of fixed variables A, B, and C from a DATA STATEMENT which follows later in the program. STATEMENT 20 instructs the computer to systematically consider the range of pressures from 0 to 500 atmospheres in steps or increments of 10 atmospheres. In the first cycle or LOOP, the pressure is set at the lower specified limit of zero atmospheres. STATEMENTS 30 and 40 are executed. STATEMENT 50 with NEXT P returns control to STATEMENT 20 where the initial pressure is incremented by 10 atmospheres. On each LOOP the NEXT P STATEMENT returns control to STATEMENT 20 where P is appropriately incremented. When the pressure reaches the maximal specified pressure of 500 atmospheres, the computer proceeds beyond the LOOP to STATEMENT 70.

1.10 READ, DATA STATEMENTS

The READ-DATA statements represent a third method by which fixed variables are assigned values. (The LET Statement and the INPUT Statement were the other two

Simple Basic Programming

methods.) In general, programs are more versatile when variables are assigned with data statements. Consider, for example, that when Program 1.9 is executed, the compressibilities of ethane are calculated. The investigator may then wish to investigate the compressibilities of another gas such as methane. The same type of polynomial may be expected to describe methane but the coefficients A, B, and C could assume different values. Program 1.9 is sufficiently general to calculate the compressibilities of methane, if DATA STATEMENT 60 is changed to describe the coefficients of methane.

READ and DATA must be used in combination with each other. A single READ statement may list a number of variables providing that they are separated by commas. DATA statements may list a number of constants separated by appropriate commas. Constants in DATA statements may be preceded by negative signs. DATA statements may appear anywhere in a BASIC program, but frequently appear at the end of the program.

1.11 FOR—NEXT STATEMENTS

Perhaps the single most useful statement in computer programming is the self-incremented loop statement which is called the FOR-NEXT statement in BASIC. This statement directs the computer to repeat a section of the program. Each time the section is repeated the value of a designated variable is incremented by a specified amount. The repetition continues until the designated variable reaches a specified value, at which time the computer exits from the loop and proceeds to the remainder of the program.

The FOR statement must be used in combination with the NEXT statement. The FOR statement defines the limits of the changing variable and specifies the desired incrementations. The FOR statement also defines the beginning of a set of statements which are to be repeated. The NEXT statement simply defines the end of the set of statements which are to be repeated. The limits for the changing variable may be specified as constants or as variables, defined or calculated elsewhere in the program. If the STEP portion of the statement is deleted, then the computer merely steps in units of one. The examples which follow illustrate several general FOR statements:

```
100 FOR X = .5 TO 100 STEP .5
100 FOR Y = 5 TO N STEP 5
100 FOR Z = M TO N STEP 2
100 FOR R = 0 TO 10 STEP J
100 FOR T = 1 TO 100
```

1.12 THE USE OF PARENTHESES

Parenthesis are used in assignment statements to group mathematical operations. When mathematical equations contain parenthesis, the same parenthesis are written into computer statements. Additional parenthesis may, however, be necessary to group computer calculations.

When a computer executes a single simple assignment statement, it systematically segments calculations. Exponential calculations are given highest priority and are, consequently, completed first. Multiplications and divisions are given second priority. They are completed in the order in which they appear. Finally the additions and subtractions are completed in the order in which they appear. Consider, for example, the sample statement

    10 LET Y = A + B*X - C/2*X + D*X↑2

The computer calculates the value of Y as defined by eight different mathematical operations. The operation with highest priority is the squaring of X. The second priority calculations evaluate the B*X, the C/2*X, and the product of D and X squared. Finally, the four terms are added or subtracted to obtain the value of Y. When parenthesis appear within an assignment statement the terms within the parenthesis are given highest priority. Study the sample statement

    20 LET Y = (A+B)*X - C/2*(X + DX↑2)

In this case, the terms within the parenthesis are evaluated first. As a result; the value of Y is defined as

$$y = (a + b)x - \frac{c}{2}(x + dx^2)$$

Parenthesis may also be nested. The terms in the inner most parenthesis are evaluated first. Therefore, a statement such as

    30 LET Y = (A + B*(X - C/2))*X + D*X↑2

has the mathematical equivalent of

$$y = [a + b(x - \frac{c}{2})]x + dx^2$$

Parenthesis must be paired. Each left parenthesis must have an associated right parenthesis.

Simple Basic Programming

Listed below are several useful mathematical equations. Sample assignment statements are suggested for each equation.

Equation	Assignment Statement
$y = x^{-2}$	LET Y = X↑(-2)
$y = x + \dfrac{a}{4}$	LET Y = X+A/4
$y = \dfrac{x + a}{4}$	LET Y = (X+A)/4
$n = \dfrac{PV}{RT}$	LET N = P*V/R/T
	LET N = P*V/(R*T)
$c = (8RT/\pi M)^{1/2}$	LET C = (8*R*T/(3.14*M))↑0.5
$E = \dfrac{h^2}{8\pi^2 I} J(J+1)$	LET E = H↑2*J*(J+1)/(8*3.14↑2*I)
$P = \dfrac{RT}{V-b} - \dfrac{a}{V^2}$	LET P = R*T/(V-B)-A/V↑2

1.13 EXERCISES

1. Consider each of the following programs. Follow the programs and complete the calculations as if you were the computer. Write out the results that the computer would print out.

    PROBLEM A:
    ```
    10 LET A = 2
    20 LET B = 3
    30 LET C = 24
    40 LET Y = C/A*B
    50 PRINT Y
    60 END
    ```

    PROBLEM B:
    ```
    10 LET X = 1
    20 LET Y = X↑2
    30 PRINT X,Y
    40 LET X = X+1
    50 IF X=11 THEN 70
    60 GO TO 20
    70 END
    ```

PROBLEM C:

```
10 READ N
20 IF N=0 THEN 70
30 LET X = N↑.5
40 PRINT N,X
50 GO TO 10
60 DATA 4,16,25,36,49,0
70 END
```

PROBLEM D:

```
10 LET X = 0
20 FOR N=1 TO 10 STEP 1
30 LET X=X+N
40 NEXT N
50 PRINT X
60 END
```

PROBLEM E:

```
10 PRINT 4↑2*3.14/8
20 END
```

2. The heat capacity, in joules per mole, for CO_2 gas is given by the polynomial

$$C_p = a + bT + cT^2$$

where $a = 26.762$
$b = 42.651 \times 10^{-3}$
$c = -147.85 \times 10^{-7}$

Write your own computer programs to calculate the heat capacity for a mole of CO_2 at 300, 400, 500, $\cdots$, 1500°K.

a) Write a program similar to the program described in Section 1.1.

b) Write a program which uses incrementation and an IF-THEN Statement as illustrated in Section 1.6.

c) Write a program which uses a FOR-NEXT loop and which uses READ-DATA Statements as illustrated in Section 1.9.

3. The heat capacity, in joules per mole, for CO gas may be expressed as:

$$C_p = a + bT + cT^{-2}$$

where $a = 28.41$
$b = 4.10 \times 10^{-3}$
$c = -0.46 \times 10^{5}$

Write a computer program to evaluate the heat capacity for a mole of CO at 300, 400, 500, $\cdots$, 2000°K.

CHAPTER 2

Using The Computer

2.1 INTRODUCTION

After a student-chemist has written his computer program, he must enter his program into the computer; he must edit his program; and finally he must instruct the computer to execute his program. If the chemist is logical and accurate in each of these steps, then the computer will return logical and accurate results. If, on the other hand, the student makes even the slightest error in any of the outlined procedures, then the results will not be satisfactory. This section outlines the basic information necessary for a student to interact successfully with the computer.

In recent years computers have been designed so that individuals could interact with a central computer via a teletype machine. In fact, a single computer may easily interact with 20 different persons on 20 separate teletypes. The computer has sufficient speed so that it can deal with each teletype operator on a time-shared basis. The successful computer calculation requires a reasonable understanding of the teletype machine and further requires a knowledge of the commands to which the computer itself will respond.

2.2 TELETYPE KEYBOARD

The teletype machine is basically a typewriter that converts typed letters into electrical signals for transmission to other teletypes or to computers. The teletype also has the capability of converting electrical impulses from a computer into typed output. Most teletypes are equipped with punch-tape attachments. When the punch attachment is activated the teletype produces a punch-tape record of computer programs, data input, and computer output. When the read attachment is activated the teletype translates the punch-tapes to electrical signals for interpretation by the computer.

The teletype keyboard consists of keys with all the capital letters of the alphabet and the arithmetic numbers from zero to 9. Special symbols are found on separate keys or as upper case keys. Some of the more useful symbols on the teletype are summarized below:

*	Multiplication of numbers
/	Division of numbers
+	Addition of numbers
-	Subtraction of numbers
-	Symbol for negative numbers
.	Decimal point
,	Comma, separation of information
>	Symbol for less than
<	Symbol for greater than
(,)	Parenthesis - for logical groupings of variables
=	Assignment operator or equality sign
break	Pressing the <u>break</u> key while a computer program is running signals the computer to terminate execution. A successful termination produces a teletype message STOP.
←	A backspace key. Each time this key is depressed, one character or space is deleted.
return	This key is used when a teletype line is completely typed. The typewriter carriage is returned to the left side of the paper. A <u>return</u> also causes the just typed line to be fed into the computer. If the computer accepts the just typed line, then the teletype advances to a new line on the typewriter. If the computer cannot decipher the typed line, then an error message will be printed on the teletype.
line feed	Advances typewriter by one line.

Using The Computer

    <u>esc</u> or <u>alt</u>-<u>mode</u> Deletes the statement or line which is being typed. The computer responds with a "/", and the carriage returns with an automatic line feed.

    <u>control</u> The <u>control</u> key is used like a shift key on a typewriter. The control key is depressed while another key is punched. The typewriter does not print letters which are punched while the control key is depressed.

    C^c The symbol means that a "C" key be depressed on the keyboard. The superscript c means that the <u>control</u> key should be depressed while typing this character. When the computer is paused and waiting for input data, the "controlled C" terminates the program.

2.3 SELECTED COMMANDS FOR THE COMPUTER

    Commands entered into the computer, via teletype, are direct instructions to the computer. They are used to gain access to the computer, to enter and manipulate programs, to execute programs, and to utilize special features of the computer system. Commands differ from program statements in that they are not preceded by statement numbers, and they are executed immediately.

    Each computer system has developed a set of commands which is specific to the system. Therefore this section can only describe the types of commands which are useful and give examples of commands used in the more popular computer systems. The potential computer user must refer to the manuals which accompany his computer system and become thoroughly familiar with the equivalent commands described in this section.

    The computer user must follow a specific procedure in gaining access to the computer. This is generally accomplished by typing a command such as HELLO followed by a user identification number and an access code. At this time the computer reserves some active memory or storage space for the user. A user may then type in his program or he may read a program from paper-punch tape. This program is stored in active memory where it is conveniently executed when a command such as RUN is typed. Data required for the execution of the current program is stored in active core; results obtained from the programmed calculation are also recorded in active memory. Specific commands will list the current program, will edit the current program or will erase part or all of the current program. Finally a command such as BYE terminates access to the computer.

    Most computers possess auxiliary memory units such as disk devices or magnetic tapes. These are particularly useful for storing frequently used programs. A command such as SAVE requests that a current program be placed into inactive memory on the auxiliary storage device. A particular terminal is usually assigned a large area on these auxiliary storage devices. The terminal eventually builds up a library of stored programs which may be recalled into active core with an appropriate command.

Finally, several commands are necessary in operating the paper-punch tape attachment on the teletypewrtier. A command such as TAPE signified that information will be transferred from tape to active computer core. A command such as PUNCH commands that the active program be recorded on paper tape.

Selected sample commands are listed below with a description of their function:

Command	Description
HELLO	Indicates that user requests access to time-sharing system. Usually followed by code or password.
GET or OLD	Requests computer to retrieve a library program from storage device and to place this program into active core.
LIST	Lists on teletypewriter the statements of the program in active core.
NAME	Informs computer that user wishes to assign a name to current program.
DELETE	Instructs the computer to remove certain statements from the current program.
SCRATCH	Erases current program from active core.
TAPE	Informs computer that information from a paper-punched tape is forthcoming.
KEY	Command used after TAPE is executed. Returns computer control to teletype keyboard.
BYE	Informs computer that present computer operator has completed his calculations.
PUNCH	Directs computer to activate teletype punch-tape attachment and to prepare a punch-tape record of the current program.
KILL or UNSAVE	Removes a designated program from program library - removes program from storage device.

Using The Computer   15

        SAVE   Places current program into program library - places program into storage device.

        RUN   Command used to execute program in active core.

        CATALOG   Requests that computer list the names of all programs which are stored in program library.

When using commands, the computer user has the option of typing the entire command or typing only the first three letters. Commands frequently require additional information such as statement numbers or program names. Sample commands listed below further illustrate the practical use of commands:

        HELLO, U210, C^CH^CD Access command followed by user identification number and access code.

        GET-PLOT   Retreives program named PLOT from library.

        NAM-ROOTS   Assigns current program the name ROOTS.

        LIS-50   Lists current program starting with statement 50.

        DEL-70, 200   Erases all current program statements between limits specified.

        LIST 100   Lists current program starting with line 100.

        UNSAVE AREA   Removes program called AREA from library.

        DELETE 70, 200   Deletes specified statements from active program.

2.4 RUNNING SAMPLE PROGRAMS

The student-chemist is now in a position to RUN his programs on the time-shared computer. It is intended that each student will use the teletype to enter the program described in Section 1.1 into the computer, that he will edit the program, and that he will execute the program to obtain useful results. In completing this task he will gain experience in the use of the teletype and the use of commands. After this first program has been successfully RUN, the student will be expected to enter and execute the programs which are described in Sections 1.6 and 1.9. Finally, the student will be in a position to RUN the programs which he has written as an exercise at the end of Chapter 1.

A sample computer RUN is presented for each of the three programs. In trying to duplicate these results the user will be required to substitute the equivalent commands which are specific for his system. For example, when the user sees a HELLO-U210, C^CH^CD, he must recognize that this is a sample access command, and he must substitute the command or procedure appropriate for his computer system. If the command KILL is used, the operator may have to substitute an UNSAVE command. Finally, the user must develop his own editing procedures. He is reminded that certain keys such as the backspace key and the escape key are useful in correcting typing errors. He is also reminded that the computer itself responds with error statements whenever it cannot decipher a statement. The more systematic "debugging" of programming errors will be discussed in Chapter 3. In the early programs, the user may assess his success by comparing his computer results to the results published with the programs.

```
HELLO-U210,D


FRIDAY    07-05-74 02:20 PM   PORT # 15
U.W. LA CROSSE WELCOMES YOU TO  L A C E  TIMESHARING.
READY


10    LET A=-.00723
20    LET B=.000015
30    LET C=4.02E-09
40    INPUT P
50    LET Z=1+A*P+B*P↑2+C*P↑3
60    PRINT Z
70    END
RUN


?10               (return)
 .929204

DONE

RUN

?20               (return)
 .861432

DONE

RUN

?30               (return)
 .796708

DONE
```

When this first simple program is RUN, the computer immediately executes statements 10, 20, and 30. When it executes statement 40, INPUT P, the teletype prints a question mark. This is an indication that the program needs some data input. The operator types in the value of pressure for the first calculation of compressibility. The computer then executes statement 50, it prints the value of

Using The Computer

compressibility, and the program ends. To evaluate the compressibility of the gas at another pressure, the program must be RUN a second time, the pressure of 20 is typed input, and the computer responds with a second compressibility. Each time the program is RUN, a new calculation is completed.

The user is encouraged to test the use of the following commands in the order indicated:

    LIST
    LIST-30
    NAME-EASY
    DELETE-50
    LIST
    CATALOG
    SCRATCH
    LIST
    BYE

The second program more clearly demonstrates the usefulness of a computer calculation. Restore access to the time-shared Basic Computer, clear the core area, type the program which is listed in Section 1.6, and RUN the program.

```
10    LET A=-.00723
20    LET B=.000015
30    LET C=4.02E-09
40    LET P=0
50    LET Z=1+A*P+B*P↑2+C*P↑3
60    PRINT P,Z
70    IF P=500 THEN 100
80    LET P=P+10
90    GOTO 50
100   END
RUN

     0              1
    10             .929204
    20             .861432
    30             .796708
    40             .735057
    50             .676503
     .               .
     .               .
     .               .
    420            .907234
    430            .984218
    440           1.06524
    450           1.15032
    460           1.23949
    470           1.33277
    480           1.43018
    490           1.53175
    500           1.6375

     DONE
```

Modify the program so that the evaluation of compressibilities begins at 300 atmospheres, [40 LET P = 300], LIST program, and RUN again.

Delete statement 70, [70, <u>return</u>], LIST program, and RUN again. When you wish to terminate execution of this program depress the teletype key labelled "break".

Clear the core area in your computer and execute the program listed in Section 1.9.

```
10   READ A,B,C
20   FOR P=0 TO 500 STEP 10
30   LET Z=1+A*P+B*P↑2+C*P↑3
40   PRINT P,Z
50   NEXT P
60   DATA -.00723,.000015,4.02E-09
70   END
RUN
```

0	1
10	.929204
20	.861432
30	.796708
40	.735057
50	.676503
.	.
.	.
.	.
420	.907234
430	.984218
440	1.06524
450	1.15032
460	1.23949
470	1.33277
480	1.43018
490	1.53175
500	1.6375

DONE

2.5 STORING PROGRAMS FOR FUTURE USE

When lengthy computer programs are used repeatedly it becomes impractical to type and edit the programs each time they are used. The user may store these programs in the computer library or he may record programs on paper-punched tape. Certain programs are sufficiently general and versatile so that they are permanent residents in a computer library. Other programs are seasonal in that they are used frequently for only a short period of time. These seasonal programs are placed into the library for the period of frequent use and are then placed on paper-punch tape for later use. Infrequently used programs should always be stored on tape in an attempt to conserve library storage space.

Using The Computer

A student-chemist may become familiar with the use of the library by completing the following exercise:

1) Enter a program into active computer core. Choose a single program from Sections 1.1, 1.6, and 1.9 and enter the program into active core.
2) LIST the program to verify the presence of this program in active core.
3) NAME the program. The program name may consist of up to six characters, and it should be an original name which has not been used by another library user. (Example: NAME-SAMPLE)
4) LIST program to verify the proper naming of the program.
5) SAVE program by typing command.
6) Type CATALOG and inspect the listing of library programs and identify your own library program.
7) SCRATCH the active core and type LIST. Notice that the program is no longer in active memory.
8) Retrieve your program from the library by an appropriate command such as GET-SAMPLE.
9) LIST program and verify its existence in active core.
10) Remove the program from the user's library by a command such as KILL-SAMPLE.
11) Type CATALOG and verify the removal of the program from the library.

A student-chemist may become familiar with the use of the paper-punch tape attachment by completing the following exercise:

1) Enter a short sample program into active computer core. LIST the program to verify its existence.
2) Activate the tape-punch attachment by depressing the ON button.
3) Type PUNCH and allow the teletype to record your program on tape.
4) De-activate the tape-punch attachment by depressing the OFF button.
5) SCRATCH the program in active memory, verify the loss of program with a LIST, and prepare to reintroduce the program via tape.
6) Type the command TAPE into the computer.
7) Thread the just prepared tape into the tape-read attachment and activate the attachment with a manual start key. As the tape is read, the teletype will print out the program.
8) When tape-input is completed, return control to the keyboard by the command KEY.
9) Verify the core existence of the program by a LIST or RUN command.

2.6 EXERCISES

1. Enter the following program into active core of the computer. Enter the program into the computer library and prepare a paper-punch record of the program.

   ```
   10 PRINT "N", "SQR. ROOT", "SQUARE", "LOG"
   20 FOR N=1 TO 100 STEP 1
   30 LET A = N↑0.5
   40 LET B = N↑2
   50 LET C = LOG(N)
   60 PRINT N,A,B,C
   70 NEXT N
   80 END
   ```

2. Determine the heat capacity of CO_2 at 300, 400, 500, $\cdots$, $1500^{\circ}K$. RUN the programs which were prepared in 1.13.2 [Chapter 1, Section 13, Problem 2.]

3. Determine the heat capacity of CO at 300, 400, 500, $\cdots$, $2000^{\circ}K$. RUN the program which was prepared in 1.13.3 [Chapter 1, Section 13, Problem 3.]

CHAPTER 3

Debugging

3.1 INTRODUCTION

The successful solution of a chemistry problem with the use of a computer is the sum of many individual steps. If each of these steps is logical and accurate, then the calculated results will be satisfactory. If the programmer makes even a single error in any of these individual steps, then a satisfactory result is not possible. It is not probable that a programmer will complete an entire problem without a single error. Errors are to be expected. The systematic search for errors and their removal is called debugging. This chapter will present some techniques to aid the programmer in preparing letter-perfect programs. A programmer, however, develops new skills as he becomes more experienced. Each discovered error will serve as a lesson in future debugging.

3.2 PROGRAM ORGANIZATION

If a computer programming problem is approached in a systematic and an organized manner, then the possibilities for errors are reduced. At the same time, a systematically organized problem is more easily debugged. An organized series of procedures for solving a computer problem will be outlined below. Each step, however, can be a potential source of error.

1. Outline the problem. Select the correct mathematical formulas for the calculation. List all of the different terms or symbols which are used in the formulas. Assign numerical values to each of the known terms. Make sure that each term is expressed in the correct units.

2. Assign variable names to each term which is used in the mathematical formulas. Select variable names for terms so that the name coincides with the symbol used for the term. For example:

    Planck's constant : H
    speed of light   : C
    gas constant   : R
    volume   : V
    pressure   : P

3. Write out the computer program on paper using BASIC symbols. Use only capital letters, use correct variable names and use computer symbols for multiplication, division, etc. Show all parentheses. Write the program in segments.

    a. Write out assignment statements for all variables which have numerical values. Write INPUT statements.
    b. Write out the major portion of the program. Set up FOR-NEXT loops, and write statements to calculate the values of unknown variables. Group calculations logically.
    c. Write subprograms in separate blocks.
    d. Write statements to PRINT the desired output.
    e. Write data statements. Place these in blocks at the beginning or the end of the program.

4. Insert REM statements to describe the program, to identify input variables, to show the position of subprograms, and to title major portions of the main program. Write PRINT statements to title output.

5. Check the written program for the following items:
    a. Parentheses in pairs - each left parenthesis is matched with a right parenthesis.
    b. Each FOR-NEXT loop has matching NEXT; FOR-NEXT loops are properly nested.

Debugging

   c. The statement number in each IF-THEN is referenced to the intended program statement.
   d. Arguements of all functions are enclosed in parentheses.
   e. Program has END statement.
   f. The program dimensions all arrays; dimension statements placed very early in program; dimensions of arrays are not exceeded.
   g. Each subprograms has a RETURN; entry to subprograms guarded by STOP, END, GO TO.

6. Type program into computer via teletype. BASIC computer systems respond to many programming and typing errors with ERROR statements. Computer ERROR statements will be discussed in Section 3.6.

7. After the program has been typed, LIST the entire program. Check each statement for typing errors. Edit and make changes where necessary. Verify the changes by frequent use of LIST.

8. Type RUN. If error messages appear continue edit phase. If numerical answers are output, attempt to verify that the results are correct.

Additional debugging techniques will be described in later sections.

3.3 AN ORGANIZED EXAMPLE

A chemist is asked to use van der Waals equation to calculate the pressures of a half mole of N_2 gas at $298.15°K$ and at volumes of 0.2, 0.4, 0.6, 0.8, $\cdots$, 2.0 liters. He is further instructed to calculate the constants in van der Waals equation from critical data. The chemist decides to write somewhat of a generalized program which will require an input temperature. The necessary equations are identified. The values of known terms are listed.

$$b = \frac{V_c}{3}$$
$$a = 3P_c V_c^2$$
$$R = \frac{8P_c V_c}{3T_c}$$
$$P = \frac{nRT}{(V - nb)} - \frac{an^2}{V^2}$$

$V_c = 0.0900$ liters
$P_c = 33.5$ atmospheres
$T_c = 126.1°K$

$n = 0.5$ moles
$T = 298.15°K$
$V = 0.2, \cdots, 2.0$ liters
$P = ?$ (atmospheres)

Computer variable names are assigned as follows:

Computer Variable Name	Term or Symbol
B	b
A	a
R	R
V1	V_c
P1	P_c
T1	T_c
P	P
V	V
T	T
N	n

The program is written in segments. The program simulates real life in that it indicates at least ten errors.

```
 10 LET    VC=.0900
 20 LET P = 33.5.
 30 LET    T1-126.1
 40 LET    N=.05
 50 INPUT T

110 LET B=(1/3*V1
120 LET A=3*P1*V1↑2
130 SET R=8*P1*V1/(3*T1)

200 FOR V=0.2 TO 2.0 STEP 0.2
210 LET P=M*R*T/(V-N*B)=A*N↑2/V↑2
220 NEXT V
215 PRINT V,P

  1 REM          VAN DER WAALS EQUATION
  2 REM     SOLVES FOR PRESSURE OF NITROGEN
  3 REM
  4 REM V1, P1, AND T1 ARE CRITICAL CONSTANTS
 49 PRINT "WHAT IS THE TEMPERATURE"
109 REM       CALCULATE V.D.W. CONSTANTS
190 REM       VAN DER WAALS EQUATION
```

The written program is checked for common errors. If none of the errors are removed before the program is typed into the computer, then the computer will respond with the following types of ERROR Statements:

MISSING ASSIGNMENT OPERATOR   IN LINE 10

CHARACTERS AFTER STATEMENT END IN LINE 20

MISSING ASSIGNMENT OPERATOR IN LINE 30

MISSING RIGHT PARENTHESIS   IN LINE 110

MISSING ASSIGNMENT OPERATOR   IN LINE 130

UNDEFINED VARIABLE IN LINE 210

LAST STATEMENT NOT 'END' IN LINE 220

Debugging

The computer calls attention to some of the errors; other errors, however, go unnoticed. The programmer must search out these errors himself. When the program is completely debugged then the LIST is

```
LIST
DEBUG

1    REM             VAN DER WAALS EQUATION
2    REM       SOLVES FOR PRESSURE OF NITROGEN
3    REM
4    REM REM    V1, P1, AND  T1 ARE CRITICAL CONSTANTS
10   LET V1=.09
20   LET P1=33.5
30   LET T1=126.1
40   LET N=.5
49   PRINT "WHAT IS THE TEMPERATURE";
50   INPUT T
109  REM      CALCULATE V.D.W. CONSTANTS
110  LET B=(1/3)*V1
120  LET A=3*P1*V1↑2
130  LET R=8*P1*V1/(3*T1)
190  REM          VAN DER WAALS EQUATION
200  FOR V=.2 TO 2 STEP .2
210  LET P=N*R*T/(V-N*B)-A*N↑2/V↑2
215  PRINT V,P
220  NEXT V
230  END
RUN
DEBUG

WHAT IS THE TEMPERATURE? 298.15
 .2              46.2898
 .4              23.416
 .6              15.6823
 .8              11.7901
 1                9.44609
 1.2              7.87965
 1.4              6.75888
 1.6              5.91726
 1.8              5.26204
 2                4.73746

DONE
```

3.4 TEST CASES

When a program does not RUN or when a program does not give numerical answers, the program operator is certainly aware that errors exist. But a program that RUNS and gives numerical answers may still contain errors. Such errors may be extremely subtle. Test cases can frequently detect the existence of these hidden errors. On the other hand, a test case can also verify that a program is functioning without error.

Consider the program which is defined in Section 3.3. A number of different test cases are possible. Examples of test cases are listed below:

1. The van der Waals pressure of a half mole of N_2 gas at 298°K when confined to one liter is known to be 9.446 atmospheres. This result can be directly compared to one of the printed computer results. If the two compare favorably, then the probability is very high that the program is without error and that the other output results are accurate.

2. A published scientific article states that the experimentally determined volume of a mole of N_2 at 50 atmospheres is .410 liters. The volume of a half mole is then .205 liters. The computed value at nearly the same pressure is close; therefore, there is a strong indication that the program is correct.

3. The ideal gas calculation for the same problem indicates that the pressure of a half mole of N_2 at 0.2 liters is 61.2 atmospheres. The computed result has the same order of magnitude and the difference is within expectations. Program is likely to be correct.

4. A colleague independently writes his own program for the same problem and obtains the same results. Probability is very high that both programs are correct.

5. The program is tested in portions. The program may be changed to set the value a = 0, the value of b = 0, and the value of R as .08206. The calculated results should coincide with the ideal gas law results. If they do, a portion of the program is quite likely to be without error.

Test cases are not always conclusive. Consider a second example where a program has been written to determine the roots of a quadratic equation by the quadratic formula. A test case may verify an accurate program for only real roots. The program may still contain errors in determining imaginary roots.

The last chapters of this book are problem chapters. In many cases the student chemist will need to write his own programs. In most cases, the answer section at the end of the book will contain test cases to aid in the debugging of programs.

3.5 TRACING

Difficult debugging problems may require tracing. Ordinarily only the final results of computer calculations are printed as output. Many intermediate calculations produce numbers which are not of interest and therefore, are not printed. But if these intermediate results are temporarily printed the source of programming

Debugging

errors may become evident. The printing of intermediate results for debugging purposes is termed tracing.

Tracing is possible by the addition of PRINT statements. If PRINT statements are inserted to output all intermediate results, then the trace is called a full trace. A full trace for the program listed in Section 3.3 would include the following additional statements:

```
 55 PRINT "THE TEMPERATURE IS";T
115 PRINT "THE B VALUE IS";B
125 PRINT "THE A VALUE IS";A
135 PRINT "THE R VALUE IS";R
205 PRINT "THE VOLUME IS";V
```

A typed RUN then outputs every intermediate result. The additional output could be difficult to analyze but the descriptive comments which have been included with each PRINT statement easily identify each printed value. The programmer then compares the computer results with results from hand calculation. The position of the first discrepancy indicates the source of an error. After the program is debugged, the extra PRINT statements may be removed.

Time sharing systems are also conducive to selective tracing. Only small portions of a program are traced during a single run. Selective tracing should be systematic. The trace may attack difficult areas first; the trace may work backward from the final result, or the trace may start from the beginning.

3.6 ERROR STATEMENTS

The computer recognizes many errors in BASIC programs and prints teletype ERROR messages to aid the program operator. The exact wording of an ERROR statement varies among computer systems, but all systems diagnose similar problems. A typical list of selected ERROR statements is presented below. A study of this list helps in the interpretation of the printed messages, but it also serves as a reference list of common errors which might be avoided. Computer manuals must be consulted for more specific descriptions of ERROR statements for specific BASIC systems.

ERROR MESSAGE	EXPLANATION OR EXAMPLE
1. CHARACTERS AFTER STATEMENT END	10 LET T=5.0./* Period, slash, and asterisk are extra characters after logical statement
2. MISSING ASSIGNMENT OPERATOR	10 LET T - 5 The equal sign is missing - replace negative with equal sign

3.	ILLEGAL EXPONENT	10 LET T=2↑-2 Write as T=2↑(-2)
4.	ILLEGAL INTEGER	L0 LET T=50 Statement number should be 10
5.	NEXT WITHOUT MATCHING FOR UNMATCHED FOR	Each FOR-NEXT loop requires corresponding NEXT
6.	UNDEFINED STATEMENT REFERENCE	100 GOTO 65 (No statement 65 in program)
7.	MISSING RIGHT PARENTHESIS MISSING LEFT PARENTHESIS	Each left parenthesis requires a matching right parenthesis
8.	UNDEFINED FUNCTION	Program contains function which is not defined
9.	FUNCTION DEFINED TWICE	
10.	BAD FUNCTION NAME	10 DEF FN2(X) Function name should not have a number in it
11.	NO CLOSING QUOTE	10 PRINT "DATA ;D Quotes should follow DATA
12.	DATA OF WRONG TYPE	The program needed a number, and operator typed a letter
13.	EXTRA INPUT-WARNING ONLY	Too many numbers are typed on an input statement. The program neglects extra input
14.	OUT OF DATA	Program executes a READ statement and DATA statements are not present or are exhausted
15.	BAD INPUT, RETYPE LAST INPUT IGNORED	Computer could not interpret data and asks operator to try again
16.	ARRAY NOT DIMENSIONED	An array was not dimensioned before it was used in program
17.	RETURN WITH NO PRIOR GOSUB	The program entered a subprogram. Guard entry with STOP
18.	GO SUBS NESTED TEN DEEP	Rewrite program with fewer GO SUBS
19.	OUT OF STORAGE	The computer library does not have sufficient space to store entire program
20.	LAST STATEMENT NOT END	Write an END statement for program
21.	LOG OF NEGATIVE ARGUEMENT	The log of a negative number cannot be defined
22.	LOG OF ZERO	The log of zero is negative infinity
23.	DIVIDE BY ZERO	The computer cannot divide by zero
24.	OVERFLOW WARNING	A variable is larger than the computer can handle, change the program

25. UNDERFLOW WARNING   The numbers are too small and below
                                       limit of computer, change the pro-
                                       gram

26. SQUARE ROOT OF NEG. ARGUEMENT   Cannot take the square root of a
                                       negative number.

CHAPTER 4

The Program Plot

4.1 INTRODUCTION

PLOT is a general "canned" program which plots a continuous function on teletypewriter paper. It is a frequently used program and should be a permanent resident of the computer library. The program, as listed in this chapter and as stored in the library, produces a plot of the sine function. By changing a single program statement, the program will plot any function where Y is a function of the X variable. The term "canned" denotes that the user need not fully understand the programming techniques utilized in the program. He need only understand the details of adopting the program to his needs.

This chapter will develop the use of functions as a more advanced technique in BASIC programming. An understanding of functions will be necessary in the successful use of the PLOT program. Specific instructions for use of this "canned" program will be followed by a LIST of the program, a sample application and a set of problems which relate to plotting in physical chemistry.

4.2 STANDARD FUNCTIONS

Calculations in physical chemistry require the use of many mathematical and trigometric functions such as log, sine, cosine, and tangent. Common functions are

The Program Plot

already defined within the computer system in terms of function statements. The programmer evaluates these functions by calling the appropriate function and by supplying the variable (argument). The types of functions defined, and the code names which are assigned, vary among computer systems. The twelve functions listed below, however, are somewhat universal.

Code	Description
SIN(X)	Gives the sine of a real argument expressed in radians.
COS(X)	Gives the cosine of a real argument expressed in radians.
TAN(X)	Gives the tangent of a real argument expressed in radians.
ATN(X)	Gives the arctangent of a real argument expressed in radians.
LOG(X)	Gives the <u>natural</u> logarithm of a positive argument.
EXP(X)	Gives value of $e \uparrow X$, where $e = 2.71828$.
ABS(X)	Gives the absolute value of the argument.
SQR(X)	Gives the square root of a positive argument.
RND(X)	Gives a random number selected between the limit of 0 and 1. X is a dummy variable and is ignored in the evaluation of the random number.
TAB(X)	Moves (tabs) teletypewriter to position specified by X.
SGN(X)	Gives a value of 1 if argument is positive; gives a value of zero if argument is zero; and gives a value of -1 if argument is negative.

Arguments in the above functions may be constants, variables, or expressions. The entire argument must be enclosed in parenthesis. Study the following examples:

```
10   LET Y = LOG (54.75)
10   LET Y = LOG (X)
10   LET Y = LOG (C-C0/C)
10   LET Y = SIN (90*3.1416/180)
10   LET R = (-B + SQR(B↑2 - 4*A*C))/(2*A)
10   LET K = A*EXP (-E/(K*T))
10   PRINT TAB(25); P; TAB(50); Z
```

Many physical chemistry texts designate natural logarithms by the ln(X) symbol and logarithms with a base ten by the log(X) symbol. The computer language uses LOG(X) as the symbol for natural logarithms. The student may obtain logarithms in base ten by dividing natural logarithms by 2.303. Trigonometric functions require argument expressed in radians. Arguments in degrees are easily converted to radians with the relationship that a full circle (360°) contains two pi radians.

4.3 DEFINED FUNCTIONS

The chemist-programmer may define additional functions within his program. Names of defined functions begin with letters FN followed by a simple variable. A single program can use up to twenty-six different functions with names FNA, FNB, FNC, FND, FNE, ... FNX, FNY, and FNZ. Functions are defined at the beginning of computer programs with special DEF programming statements which are illustrated below:

```
10   DEF FNY(X) = X↑2 + 2*X + 6
10   DEF FNZ(P) = 1 + A*P + B*P↑2 + C*P↑3
10   DEF FNV(T) = n*R*T/P
```

DEF is the defining operator, the function code includes the name with the argument in parentheses, and a defining expression follows the equal sign. Variables in the defining expression are presumably defined in the remainder of the program. Newly defined functions may be defined in terms of previously defined functions.

The use of a function within a program may be demonstrated by a simple example. Consider the use of the ideal gas equation in calculating the volume of a mole of gas at room temperature over a 1-10 atmosphere pressure range.

```
10   DEF FNV(P) = n*R*T/P
20   LET n = 1
30   LET R = .082
40   LET T = 300
50   FOR P = 1 TO 10
60   PRINT P, FNV(P)
70   NEXT P
80   END
```

4.4 USING THE PLOT PROGRAM

The PLOT program as listed produces a graph of the sample function which is described by statement 500.

```
500   DEF FNF(X) = SIN(X)
```

The Program Plot

When the command RUN is typed, the computer will request additional teletype information from the user. The user identifies the plotting interval by inputting the value of X where the plot is intended to start, and the value of X where the plot is intended to terminate. He will also specify the total number of points to be plotted over the interval. The computer program will automatically select regularly spaced values of X over the described interval, it will calculate the values of Y at each of these selected points, it will scale the plot in the Y direction so that it fits on the teletypewriter paper, and it will produce a properly labeled plot. The X-axis of the plot lies parallel to the teletype paper as it is fed into the machine. The program operator also has the additional option of listing all of the X-Y data pairs which were used in the plot.

When the program operator wishes to plot a new function he may use the same program but he must replace program statement 500 with a new statement 500 which defines the new function. The argument or the independent variable must be expressed in terms of the variable X. The operator may define parameters needed in the function statement by statements with numbers below 500. All of the simple variables (A,B,C....X,Y,Z) have been reserved for this purpose.

Instructions for the use of PLOT are outlined below with sample input:

1. Call program from library.
    GET-PLOT

2. Define variables used in function definition using only simple variables.
    470   LET A = −.00723
    480   LET B = .000015
    490   LET C = 4.02E−09

3. Define new function statement 500 in terms of X argument.
    500   DEF FNF(X) = 1 + A*X + B*X↑2 + C*X↑3

4. Type RUN command and answer following questions when requested by teletype printer:

    a. Value of X where plotting interval begins?

    b. Value of X where plotting interval ends?

    c. Total number of points to be plotted?

    d. Do you want a listing of plotted values?   (Yes, No)

LIST
PLOT

```
500   DEF FNF(X)=SIN(X)
510   DIM Z[1000],P[4]
520   PRINT "WHERE DO YOU WISH TO BEGIN THE"
530   PRINT "EVALUATION OF THE CONTINOUS FUNCTION";
540   INPUT X1
550   PRINT
560   PRINT "AT WHICH POINT DO YOU WISH TO STOP THE EVALUATION";
570   INPUT X2
580   PRINT
590   PRINT "HOW MANY SEGMENTS IN THE X-INTERVAL";
600   INPUT N9
610   PRINT
620   PRINT "DO YOU WANT A LISTING OF THE VALUES, (1=YES,0=NO)";
630   INPUT G1
640   PRINT
650   PRINT "X-VALUE";TAB(25);"*** Y - AXIS ***"
660   PRINT " BELOW"
670   M1=-1.E+37
680   M2=+1.E+37
690   D1=(X2-X1)/N9
700   FOR I9=1 TO N9+1
710   X3=X1+(I9-1)*D1
720   Z[I9]=FNF(X3)
730   IF Z[I9] >= M2 THEN 750
740   M2=Z[I9]
750   IF Z[I9] <= M1 THEN 770
760   M1=Z[I9]
770   NEXT I9
780   P[1]=M2
790   P[2]=M2+18*(M1-M2)/56
800   P[3]=M2+36*(M1-M2)/56
810   P[4]=M2+54*(M1-M2)/56
820   PRINT TAB(10);P[1];TAB(27);P[2];TAB(43);P[3];TAB(60);P[4]
830   FOR I9=1 TO 72
840   IF I9=16 THEN 900
850   IF I9=34 THEN 900
860   IF I9=52 THEN 900
870   IF I9=68 THEN 900
880   PRINT "-";
890   GOTO 910
900   PRINT "Y";
910   NEXT I9
920   Z1=INT(-M2*56/(M1-M2)+1.5)+15
930   FOR I9=1 TO N9+1
940   S2=X1+(I9-1)*D1
950   PRINT S2;TAB(15);
960   Y1=INT(56*(Z[I9]-M2)/(M1-M2)+1.5)+15
970   T1=ABS(S2)-D1*.5
980   FOR J9=16 TO 72
990   IF Y1=J9 THEN 1070
1000  IF Z1=J9 THEN 1090
1010  IF T1<0 THEN 1110
1020  IF J9 <= Y1 THEN 1050
1030  IF J9 <= Z1 THEN 1050
```

The Program Plot

```
1040    GOTO 1140
1050    PRINT " ";
1060    GOTO 1120
1070    PRINT "+";
1080    GOTO 1120
1090    PRINT "O";
1100    GOTO 1120
1110    PRINT "Y";
1120    NEXT J9
1130    GOTO 1150
1140    PRINT
1150    NEXT I9
1160    IF G1=0 THEN 1250
1170    FOR I9=1 TO 5
1180    PRINT
1190    NEXT I9
1200    PRINT "X-VALUE","Y-VALUE"
1210    FOR I9=1 TO N9+1
1220    X3=X1+(I9-1)*D1
1230    PRINT X3,Z[I9]
1240    NEXT I9
1250    END


GET-PLOT
RUN
PLOT

WHERE DO YOU WISH TO BEGIN THE
EVALUATION OF THE CONTINOUS FUNCTION? 0

AT WHICH POINT DO YOU WISH TO STOP THE EVALUATION? 6.2832

HOW MANY SEGMENTS IN THE X-INTERVAL? 15

DO YOU WANT A LISTING OF THE VALUES, (1=YES,0=NO)? 1

X-VALUE                     *** Y - AXIS ***
  BELOW
              -.994523         -.355187          .284148           .923484
---------------Y-----------------Y-----------------Y---------------Y----
0              YYYYYYYYYYYYYYYYYYYYYYYYYYYYY+YYYYYYYYYYYYYYYYYYYYYYYYYYYYY
 .41888                                         0        +
 .83776                                         0              +
1.25664                                         0                        +
1.67552                                         0                        +
2.0944                                          0              +
2.51328                                         0     +
2.93216                                         0 +
3.35104                           +             0
3.76992                      +                  0
4.1888                  +                       0
4.60768          +                              0
5.02656            +                            0
5.44544                 +                       0
5.86432                            +            0
6.2832                                          +
```

X-VALUE	Y-VALUE
0	0
.41888	.406738
.83776	.743146
1.25664	.951058
1.67552	.994521
2.0944	.866023
2.51328	.58778
2.93216	.207905
3.35104	-.20792
3.76992	-.587793
4.1888	-.866031
4.60768	-.994523
5.02656	-.951053
5.44544	-.743136
5.86432	-.406723
6.2832	1.49803E-05

DONE

4.5 SAMPLE CALCULATION

Consider again the familiar example from Chapter 1 where the compressibilities for ethane are described by the function

$$Z = 1 + AP + BP^2 + CP^3$$

where the coefficients at room temperature assume the following values:

$A = -7.23\ E-03$

$B = 1.5\ E-05$

$C = 4.02\ E-09$

The PLOT program, if properly modified, will produce a graphical display of this function and will also output the calculated values of compressibility at regular intervals of pressure. Study the following computer output as a sample calculation using the PLOT program:

```
GET-PLOT
470 LET A = -7.23 E-03
480 LET B = 1.50 E-05
490 LET C = 4.02E-09
500 DEF FNF(X) = 1 + A*X + B*X↑2 + C*X↑3

RUN
PLOT

WHERE DO YOU WISH TO BEGIN THE
EVALUATION OF THE CONTINOUS FUNCTION? 0
```

The Program Plot 37

AT WHICH POINT DO YOU WISH TO STOP THE EVALUATION? 500

HOW MANY SEGMENTS IN THE X-INTERVAL? 20

DO YOU WANT A LISTING OF THE VALUES, (1=YES,0=NO)? 1

```
X-VALUE                *** Y - AXIS ***
 BELOW
           .178415         .647407         1.1164          1.58539
    ---------------Y-----------------Y-----------------Y----------------Y----
  0              YYYYYYYYYYYYYYYYYYYYYYYYYYYYYYY+YYYYYYYYYYYYYYYYYYYYYYYYYYY
 25                                           +
 50                                 +
 75                           +
100                     +
125               +
150            +
175          +
200        +
225        +
250        +
275         +
300            +
325              +
350                 +
375                    +
400                         +
425                             +
450                                  +
475                                         +
500                                                              +
```

```
X-VALUE         Y-VALUE
   0             1
  25             .828688
  50             .676503
  75             .543821
 100             .43102
 125             .338477
 150             .266568
 175             .21567
 200             .18616
 225             .178415
 250             .192813
 275             .229728
 300             .28954
 325             .372624
 350             .479357
 375             .610117
 400             .76528
 425             .945223
 450            1.15032
 475            1.38096
 500            1.6375
```

DONE

4.6 EXERCISES

1. The heat capacity, in joules per mole, for CO_2 gas is given by the polynomial
$$C_p = a + bT + cT^2$$
where $a = 26.762$
$b = 42.651 \times 10^{-3}$
$c = 147.90 \times 10^{-7}$
PLOT the heat capacity of CO_2 from $300°K$ to $1500°K$.

2. The heat capacity, in joules per mole, for CO gas may be expressed as
$$C_p = a + bT + cT^{-2}$$
where $a = 28.41$
$b = 4.10 \times 10^{-3}$
$c = -0.46 \times 10^5$
PLOT the heat capacity of CO from $300°$ to $2000°K$.

3. The functions listed below find frequent use in physical chemistry:

$$f_1(x) = \frac{x}{e^x - 1} \quad (1)$$

$$f_2(x) = \frac{x^2 e^x}{(e^x - 1)^2} \quad (2)$$

$$f_3(x) = -\ln(1 - e^{-x}) \quad (3)$$

PLOT each of these functions from $x = 0.001$ to $x = 5.001$. The computer cannot evaluate these functions at $x = 0$.

CHAPTER 5

The Program Area

5.1 DESCRIPTIVE COMMENTS WITHIN PROGRAMS

The computer program AREA, like most other programs, is much simpler to use when it contains descriptive comments. A descriptive comment may be so elaborate as to contain several paragraphs of narrative or simply a single word as a heading for column output. A descriptive comment may identify a variable or ask a worded question when data is needed.

An unfamiliar program is listed below, with a sample RUN. The program is itself quite simple and it is relatively short. The program appears somewhat larger, however, because it contains REMARK statements and descriptive PRINT statements. Study the program, follow the progress of the RUN and try to grasp what the program is doing. At the same time, be aware that descriptive comments are simplifying the use of this program.

```
LIST
COMNT

10    REM           THIS PROGRAM USES THE IDEAL GAS EQUATION
20    REM
30    REM                      P * V = N * R * T
40    REM
50    REM           TO CALCULATE THE VOLUME OF A GAS AT VARIOUS
60    REM           PRESSURES AND AT A SPECIFIED TEMPERATURE
```

```
70   REM
80   REM
90   LET R=.08206
100  PRINT "WHAT IS THE TEMPERATURE IN DEGREES CENTIGRADE";
110  INPUT T
120  PRINT "WHAT IS THE NUMBER OF MOLES";
130  INPUT N
140  PRINT "THE ABSOLUTE TEMPERATURE IS";
150  LET T=T+273.15
160  PRINT T
170  PRINT "THE GAS CONSTANT IS";R
180  PRINT
190  PRINT
200  PRINT "PRESSURE","VOLUME"
210  FOR P=1 TO 10 STEP 1
220  LET V=N*R*T/P
230  PRINT P,V
240  NEXT P
250  END
RUN
DEBUG

WHAT IS THE TEMPERATURE IN DEGREES CENTIGRADE? 100
WHAT IS THE NUMBER OF MOLES? 2.4
THE ABSOLUTE TEMPERATURE IS 373.15
THE GAS CONSTANT IS .08206

PRESSURE         VOLUME
 1                73.4897
 2                36.7448
 3                24.4966
 4                18.3724
 5                14.6979
 6                12.2483
 7                10.4985
 8                 9.18621
 9                 8.16552
10                 7.34897

DONE
```

Remark statements, which are abbreviated as REM, are computer statements which contain descriptive comments of the computer program itself. The REM lines are part of a BASIC program and therefore, they are printed when the program is listed, they are stored in the program library, and they are recorded on paper punch tape. They are essentially an attached label. REM statements, however, are not executed; they are completely ignored during a program RUN. In the above program the first eight lines illustrate the use of REM statements.

Comments within the computer output are introduced with PRINT statements. Any information which is placed between "quotes" will be printed when the PRINT statement is executed. A single PRINT statement may contain comments plus instructions to output the values of variables. Consider statement 170 in the above program. When this line is executed, the words in quotes are printed followed by the printed value of R. The semicolon instructs the computer to type the value of R immediately following the comment. Consider next, the statement 140. When this line is executed, the words in quotes are printed. The semicolon indicates that some additional printing should follow immediately. Statement 150 calculates the absolute value of T and statement 160 instructs that T should be printed. The T however, is printed on the same teletype lines that statement 140 began.

Consider also statement 200 and statement 230. Statement 230 will produce a list of P's and V's; statement 200 prints titles at the beginning of the list. When comments or variables in PRINT statements are separated by commas, the words and numbers are automatically placed into columns which are approximately fourteen spaces wide. Therefore statement 200 prints the title PRESSURE over the first column, and prints the title VOLUME over the second column. The absence of a comma or a semicolon at the end of statement 200 signifies that the printing of the teletype line is complete and activates the carriage return on the printer. Statement 230 prints the values of P and V on a single line each time it is executed.

Finally, consider how INPUT statement 130, which ordinarily produces a simple question mark, is converted into a worded question. The preceeding statement 120 prints the words, WHAT IS THE NUMBER OF MOLES. Since statement 120 is followed by a semicolon, the teletypewriter will proceed with additional printing on the same teletype line. The execution of statement 130 places a question mark at the end of the worded question. The formation of a teletype printed question is complete.

5.2 DESCRIPTION OF AREA

The program entitled AREA numerically integrates a mathematical function. The integration of a real function is equivalent to finding the area under the curve. Consequently, the logical name for the program is AREA. The mathematical description of the program which follows will utilize a practical example from thermodynamics.

The heat capacity of any substance is found to be dependent on the temperature. The heat capacities of certain substances can be described by an equation such as

$$C_p = a + bT + cT^2 \qquad (1)$$

where C_p is the heat capacity in calories per mole, T is absolute temperature and a, b, and c are coefficients which assume different values for different substances.

The heat required to increase the temperature of a mole of this substance from a temperature T_1 to a higher temperature T_2 is given by

$$\text{Heat} = \int_{T_1}^{T_2} (a + bT + cT^2) \, dT \qquad (2)$$

The heat required is determined by integrating the heat capacity function over the interval T_1 to T_2. The integral might be evaluated by integrating each term by normal calculus procedures to obtain

$$\text{Heat} = a(T_2 - T_1) + \frac{b}{2}(T_2^2 - T_1^2) + \frac{c}{3}(T_2^3 - T_1^3) \qquad (3)$$

and by simple substitution of T_1 and T_2 to yield the amount of heat required.

Consider evaluating the same integral as expressed in equation (2) by numerical techniques. The plot of the heat capacity function is shown in Figure 5.1. The area under the curve between temperature T_1 and temperature T_2 has the same

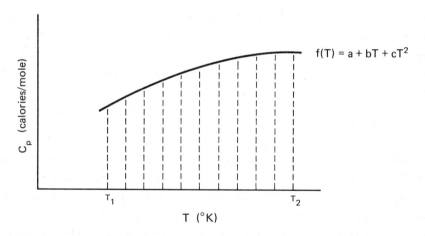

Figure 5.1 Plot of heat capacities as a function of temperature

numerical answer as "Heat" in equation (3). The computer determines the area under the curve by first dividing the interval, T_1 to T_2, into smaller intervals. This is illustrated in Figure 5.1. The total area is now the sum of the ten smaller areas. In a very simple approach, the area of each segment is approximately the width of the segment multiplied by the average height of the segment. It should be evident that some error is introduced when the segments are pictured as trapazoids. Nevertheless, the error would be reduced if the same T_1 to T_2 interval were divided into a 100 or a 1000 intervals. In general, as the number of subdivisions increases, the calculated area becomes more accurate and finally converges to the "Heat" as calculated in equation (3).

The Program Area 43

The program AREA evaluates the area under the curve by Simpson's rule, which is frequently more accurate and which frequently requires fewer intervals then the trapazoidal rule. When using Simpson's rule, the T_1 to T_2 interval is again divided into smaller intervals, but the upper boundary of each segment is treated as a parabola with the subsequent determination of each segment area. The sum of the segment areas yields the total area. Additional information concerning Simpson's rule is available in most textbooks dealing with numerical analysis.

5.3   USING THE PROGRAM AREA

The AREA program is a frequently used program and should be a permanent resident of the computer library. The program AREA is interactive in nature in that it will request information from the user. It will ask the user to define the total interval and the number of segments to be used in Simpson's technique. It will then calculate the area. The user will repeat the same calculation with a larger number of segments and compare the computed area with the area from the previous calculation. If the two areas agree, then the integration is complete and accurate. If they do not agree, another calculation with more segments is required. This procedure is repeated until a converged area is obtained.

The program, as stored on the computer, integrates the function as defined in statement 500.

```
500   DEF FNF(X) = X
```

When the program user wishes to integrate a new function, he may use the same program but he must replace the program statement 500 with a new statement 500, which defines the new function. The argument or the independent variable must be expressed in terms of the variable X. The operator may define parameters needed in the function statement by statements with numbers below 500. All the simple variables (A, B, C, $\cdots$, X, Y, Z) have been reserved for this purpose.

Instructions for the use of AREA are outlined below with sample input illustrated:

1. Call program from library.
   ```
   GET-AREA
   ```
2. Define variables used in the definition of the new function, using only simple variables.
   ```
   470   LET A = 6.214
   480   LET B = 10.396E-03
   490   LET C = -35.45E-07
   ```

3. Define new function statement 500 in terms of the X argument.

    500   DEF FNF(X) = A + B*X + C*X↑2

4. Type RUN command and answer following questions when listed on teletype printer:

    a. Value of X where integration is to begin?

    b. Value of X where integration is to end?

    c. Number of segments to be used in Simpson's rule?

5. The computer will calculate a value of the integral, and will return with the question:

    a. Do you wish to use more segments? (Yes, No)

       Check convergence of the integral by answering "yes" and responding to the question:

    b. Number of segments to be used in Simpson's rule?

6. The computer will calculate another value for the integral. Compare this value with value from previous calculation. If the values agree, you have a converged answer. If the values do not agree, repeat Item (5).

LIST
AREA

```
500    DEF FNF(X)=X
510    PRINT "WHAT IS THE LOWER LIMIT OF INTEGRATION";
520    INPUT S1
530    PRINT
540    PRINT "WHAT IS THE UPPER LIMIT OF INTEGRATION";
550    INPUT S2
560    PRINT
570    PRINT "HOW MANY SEGMENTS IN THE INTERVAL";
580    INPUT N9
590    PRINT
600    IF N9 <= 2 THEN 630
610    IF INT(N9/2)*2#N9 THEN 640
620    GOTO 660
630    PRINT
640    PRINT "THIS MUST BE A NUMBER GREATER THAN 2 AND EVEN."
650    GOTO 560
660    PRINT "THE VALUE OF THE INTEGRAL IS";
670    H8=(S2-S1)/N9
680    S3=0
690    S4=0
700    FOR I9=2 TO N9-2 STEP 2
710    S3=S3+FNF(S1+(I9-1)*H8)
720    S4=S4+FNF(S1+I9*H8)
730    NEXT I9
740    H9=H8/3*(4*(S3+FNF(S1+(N9-1)*H8))+2*S4+FNF(S1)+FNF(S2))
750    PRINT H9
760    FOR I9=1 TO 4
770    PRINT
780    NEXT I9
```

```
790  PRINT "DO YOU WISH TO USE MORE SEGMENTS, (1=YES,0=NO)";
800  INPUT D9
810  IF D9=1 THEN 560
820  END
```

5.4 SAMPLE CALCULATION

The heat capacity for carbon dioxide is given by

$$C_p = a + bT + cT^2$$

where T is absolute temperature and the coefficients are:

$a = 6.214$
$b = 10.396 \times 10^{-3}$
$c = -35.45 \times 10^{-7}$

Calculate the heat required to increase the temperature of the carbon dioxide from $298.15°K$ to $1000°K$ using

$$\text{heat required} = \int_{T_1}^{T_2} (a + bT + cT^2) \, dT$$

The AREA program, if properly modified will calculate the value of the integral. Study the following computer output as a sample calculation using the AREA program:

```
GET-AREA

470 LET A = 6.214
480 LET B = 10.396 E-03
490 LET C = -35.45 E-07
500 DEF FNF(X) = A + B*X + C*X↑2
RUN
AREA

WHAT IS THE LOWER LIMIT OF INTEGRATION? 298.15

WHAT IS THE UPPER LIMIT OF INTEGRATION? 1000

HOW MANY SEGMENTS IN THE INTERVAL? 8

THE VALUE OF THE INTEGRAL IS 7946.88


DO YOU WISH TO USE MORE SEGMENTS, (1=YES,0=NO)? 1

HOW MANY SEGMENTS IN THE INTERVAL? 20

THE VALUE OF THE INTEGRAL IS 7946.88


DO YOU WISH TO USE MORE SEGMENTS, (1=YES,0=NO)? 0

DONE
```

The first attempt at finding the value of the integral used 8 segments and produced an area of 7946.88. The second attempt at finding a refined value of the integral used 20 segments and produced an area of 7946.88. The calculation has converged and the heat required is 7946.88 calories. The user is advised against the use of excessive segment numbers to avoid round-off error. Areas are probably accurate to five significant numbers.

5.5 DIFFICULTIES IN USING AREA

Illustrations will be given for two situations which may present difficulties in computing the value of an integral. Procedures for dealing with the situations will also be outlined. Consider the following function

$$f(x) = \frac{x}{e^x - 1} \tag{4}$$

The plot of equation (4) is shown in Figure 5.2.

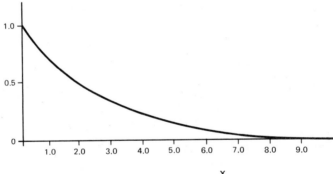

Figure 5.2 Graph of $\dfrac{x}{e^x - 1}$

The first type of problem arises when the integral, which is defined below, is attempted by computer.

$$\text{AREA} = \int_0^1 \frac{x}{e^x - 1} \, dx \tag{5}$$

When the computer is evaluating this integral, the denominator becomes so small as x approaches zero that it exceeds the capability of the machine. Remember that many computers cannot handle numbers which are less than 10^{-38}. If x equals zero, then the computer is asked to divide by zero, a situation which may cause the program to terminate. An approximate value of the same integral may be obtained by completing the integration from a very small value of x which is nearly zero. If this x value is sufficiently small, then the error in evaluating the integral is negligible. Study the following computer output which evaluates equation (5) by the procedures just described.

```
GET-AREA

500 DEF FNF(X) = X/(EXP(X)-1)
RUN
AREA
WHAT IS THE LOWER LIMIT OF INTEGRATION? .00001

WHAT IS THE UPPER LIMIT OF INTEGRATION? 1

HOW MANY SEGMENTS IN THE INTERVAL? 300

THE VALUE OF THE INTEGRAL IS .777493



DO YOU WISH TO USE MORE SEGMENTS, (1=YES,0=NO)? 0

DONE

RUN
AREA
WHAT IS THE LOWER LIMIT OF INTEGRATION? .000001

WHAT IS THE UPPER LIMIT OF INTEGRATION? 1

HOW MANY SEGMENTS IN THE INTERVAL? 300

THE VALUE OF THE INTEGRAL IS .777557



DO YOU WISH TO USE MORE SEGMENTS, (1=YES,0=NO)? 0

DONE
```

When the integration begins at x = .00001, and the number of segments is set very high, the value of the integral is 0.7775. The last digits are not significant because of round-off errors. When the integration begins at x = .000001 and the number of segments is sufficiently high, the value of the integral is 0.7775. The integral in equation (5) has the converged area of 0.7775. The error due to the shortened interval is approximately .000001 and is certainly negligible.

A second type of problem arises when the integral, which is defined below, is attempted by computer.

$$\text{AREA} = \int_1^\infty \frac{x}{e^x - 1}\, dx \qquad (6)$$

Inspection of Figure 5.2 shows that the area at large values of x becomes extremely small. The evaluation of the integral to a large value of x yields essentially the same result as integrating to an infinite value of x. If the program operator, however, selects a value of x which is too large, the contributing area for some segments will fall below the 10^{-38} limit of the computer, and the computer evaluation fails. An experienced program operator will slowly increase the upper integration limit until the value of AREA converges. Study the following computer output which evaluates equation (6) by the procedure just described.

```
RUN
AREA

WHAT IS THE LOWER LIMIT OF INTEGRATION? 1

WHAT IS THE UPPER LIMIT OF INTEGRATION? 10

HOW MANY SEGMENTS IN THE INTERVAL? 300

THE VALUE OF THE INTEGRAL IS .86693

RUN
AREA

WHAT IS THE LOWER LIMIT OF INTEGRATION? 1

WHAT IS THE UPPER LIMIT OF INTEGRATION? 20

HOW MANY SEGMENTS IN THE INTERVAL? 300

THE VALUE OF THE INTEGRAL IS .867429

RUN
AREA

WHAT IS THE LOWER LIMIT OF INTEGRATION? 1

WHAT IS THE UPPER LIMIT OF INTEGRATION? 30

HOW MANY SEGMENTS IN THE INTERVAL? 300

THE VALUE OF THE INTEGRAL IS .86743
```

In each case, the number of segments is set sufficiently large to insure convergence due to Simpson's rule. When the integration is completed to x = 10, the value of the integral .86693. Integration to large values of x = 20 or x = 30 yields essentially the same result of .86743. The value of the integral described by equation (6) is, therefore, .86743.

5.6 EXERCISES

1. Calculate the heat in joules which is required to increase the temperature of a mole of CO gas from 300 to $1000°K$. The heat required is expressed as ΔH and is evaluated by integrating the heat capacity function as follows:

$$\Delta H = \int (a + bT + cT^{-2}) \, dT$$

where $a = 28.41$
$b = 4.10 \times 10^{-3}$
$c = -0.46 \times 10^{5}$

2. The wave function for a particle in a one-dimensional box is given by

$$\psi(x) = \sqrt{2/a} \, \sin \frac{n\pi x}{a}$$

where a is the length of the box and where n is a quantum number. The probability of finding the particle somewhere between $x = 0$ and $x = a/4$ is given by

$$\text{Prob.} = \int_0^{a/4} (2/a) \left(\sin \frac{n\pi x}{a}\right)^2 dx$$

Given an electron in a 10 Å box which is in the n = 1 state. Calculate the probability of finding the electron somewhere between $x = 0$ and $x = a/4$.

3. Determine the values of each of the following definite integrals:

a. $\int_0^{\pi} x^2 \cos^2 x \, dx$

b. $\int_0^{\pi} \frac{\sin x}{x} dx$

c. $\int_0^{\infty} e^{-2x} dx$

d. $\int_0^{\infty} xe^{-x^2} dx$

CHAPTER 6

The Program LineQ

6.1 DESCRIPTION OF LINEQ

The program LINEQ fits a set of plotted x and y data points with a best line by the least squares method. It determines the slope and intercept of the best line. It also aids the operator in judging the accuracy of the line fit by determining the standard deviations of the slope and intercept, the correlation coefficient for the line fit, and an appropriate difference table. The program LINEQ can manipulate seven different linear equation types, sufficient to handle most applications in physical chemistry. LINEQ is a frequently used program and should be a permanent resident in a computer library.

A short description of the method of least squares will follow. Consider the following x-y points as data for a line fit:

$x_1 = 0$ $y_1 = 1.1$

$x_2 = 1$ $y_2 = 1.9$

$x_3 = 2$ $y_3 = 3.0$

$x_4 = 3$ $y_4 = 4.1$

$x_5 = 4$ $y_5 = 5.0$

$$x_6 = \qquad y_6 = 4.9$$
$$\vdots \qquad \vdots$$
$$x_n \qquad y_n$$

It is evident by simple inspection that the data can be described (fit) by a straight line,

$$y = mx + b \qquad (1)$$

Since the data points have experimental scatter, a line cannot possibly be drawn through all of the data points. Therefore, a number of different lines are proposed and each line is described by a different equation. Each proposed equation for the line fit has a slightly different slope and intercept. The problem is to select the best line which is described by a unique slope and intercept. Even when the best line is considered, there are differences between the plotted points and the points on the line. These differences are termed residues and are calculated as follows:

$$r = y_{calc} - y_{exp} \qquad (2)$$

where r is the residue of a single data point, where y_{calc} is a y value from Equation (1) and y_{exp} is a data point. In the least squares technique, the residue for each data point is calculated, each residue is squared, and the squared residues are summed.

$$S = \sum_{i=1}^{n} (x_i m + b - y_i)^2 \qquad (3)$$

where the summation is complete over all n data points, where the $(x_i m + b)$ is a calculated y, and where y_i is an experimental data point. Each proposed line has a different S, the best line would be the one that has a minimum value of S.

Since S is a function of both m and b, the conditions for a minimum S must be

$$\frac{\delta S}{\delta m} = 0 = 2 \sum_{i=1}^{n} x_i (x_i m + b - y_i) \qquad (4)$$

$$\frac{\delta S}{\delta b} = 0 = 2 \sum_{i=1}^{n} (x_i m + b - y_i) \qquad (5)$$

Equations (4) and (5) are then solved for the best value of m and b,

$$m = \frac{n \Sigma y_i x_i - \Sigma x_i \Sigma y_i}{n \Sigma x_i^2 - (\Sigma x_i)^2} \qquad (6)$$

$$b = \frac{\Sigma x_i^2 \Sigma y_i - \Sigma x_i \Sigma x_i y_i}{n \Sigma x_i^2 - (\Sigma x_i)^2} \qquad (7)$$

The slope and intercept are entirely described by summations involving the experimental x and y data points. Remember that n is the number of experimental data pairs.

The standard deviations for the slope and intercept may be calculated as follows:

$$\pm \delta_m = \left[\frac{n}{n-2} \cdot \frac{\Sigma r_i^2}{(\Sigma x_i)^2 - n \Sigma x_i^2} \right]^{1/2}$$

$$\pm \delta_b = \left[\frac{1}{n-2} \cdot \frac{\Sigma r_i^2 \Sigma x_i^2}{(\Sigma x_i)^2 - n \Sigma x_i^2} \right]^{1/2}$$

where the r_i values are residues as previously defined. The correlation coefficient is defined by

$$C = \frac{n \Sigma x_i y_i - \Sigma x_i \Sigma y_i}{[(n \Sigma x_i^2 - (\Sigma x_i)^2)(n \Sigma y_i^2 - (\Sigma y_i)^2)]^{1/2}}$$

LINEQ provides for the use of seven different types of equations. Data points for each of the equations are indicated by X and Y. The computer must manipulate some of the data before it fits the general form of a straight line as described in Equation (1). Each of the seven equation options are listed in the left hand column. The computer calculations, which are necessary to place the data into linear form, are indicated in the right hand column.

Equation Type	Preliminary Computer Calculations
1. Y = M*X + B	none necessary
2. Y + M*(1/X) + B	x = (1/X)
3. LOG(Y) = M*X + B	y = LOG(Y)
4. LOG Y = M*(1/X) + B	y = LOG(Y)
	x = (1/X)
5. (1/Y) + M*X + B	y = (1/Y)
6. LOG(Y) = M*LOG(X) + B	y = LOG(Y)
	x = LOG(X)
7. Y = M*LOG(X) + B	x = LOG(X)

LINEQ has appropriate sub-programs to complete each of the preliminary calculations.

6.2 NUMERIC DATA IN ARRAYS

A program such as LINEQ uses a large number of x-y data points. It is difficult to assign a different variable to each x and y point. It is desirable to identify each x data point with the variable X, and to use subscripts with the X to identify each separate variable. Each y, then, would be identified by the variable Y, and subscripts with the variable Y would identify specific y data points. The use of subscripted variables is further illustrated in the list which follows:

$x_1 = 0$	X(1)	$y_1 = 1.1$	Y(1)
$x_2 = 1$	X(2)	$y_2 = 1.9$	Y(2)
$x_3 = 2$	X(3)	$y_3 = 3.0$	Y(3)
$x_4 = 3$	X(4)	$y_4 = 4.1$	Y(4)
$x_5 = 4$	X(5)	$y_5 = 4.9$	Y(5)
$x_6 = 5$	X(6)	$y_6 = 6.0$	Y(6)
⋮	⋮	⋮	⋮

The input data point x_1 is named variable X(1), input data point x_2 is named variable X(2), etc. The x data points are in an ordered collection which is called an array. The y data points are in a second ordered array, where individual data points are identified as Y(1), Y(2), Y(3), $\cdots$, Y(N) where there are N data points.

Arrays must be stored in blocks within the computer system. The computer programmer must reserve these blocks of space within the computer before any reference is made to the subscripted variables. A DIMENSION STATEMENT, which appropriates the space for arrays, has the following form:

    500 DIM X(100), Y(100)

This statement allocates space for up to a 100 x-variables and up to a 100 y-variables. All the positions in a dimensional array need not be used. Arrays are dimensioned sufficiently large to avoid problems.

A short demonstration program illustrates the use of arrays. The program will read 6 x-y data points; it will calculate the product of each of the x-y data points and store each product term in an array Z; and it will also sum the points in each array.

```
LIST
ARRAY

10    DIM X[20],Y[20],Z[20]
20    READ N
30    FOR I=1 TO N
40    READ X[I],Y[I]
50    NEXT I
60    FOR J=1 TO N
70    LET Z[J]=X[J]*Y[J]
80    NEXT J
90    LET X=0
100   LET Y=0
110   LET Z=0
120   FOR K=1 TO N
130   LET X=X+X[K]
140   LET Y=Y+Y[K]
150   LET Z=Z+Z[K]
160   NEXT K
170   PRINT X,Y,Z
180   DATA 6
190   DATA 0,1.1
200   DATA 1,1.9
210   DATA 2,3
220   DATA 3,4.1
230   DATA 4,4.9
240   DATA 5,6
250   END
RUN
ARRAY

 15            21            69.8

DONE
```

The Program LineQ

The first DATA line will inform the computer that six sets of data are to be expected. Six additional DATA lines identify the data pairs. The use of arrays practically implies the use of FOR-NEXT statements. Each major operation in the program utilizes a FOR-NEXT loop from 1 to N where N is defined by the first DATA statement. The FOR-NEXT loop in statement 30, reads the data statements and fills the six positions in the X and Y arrays. The FOR-NEXT loop in statement 60 uses data from the X and Y arrays to calculate the pair products and enters the results into the first six positions of the Z array. The FOR-NEXT loop in statement 120 sums the X, Y, and Z Arrays. Notice that the variables X, Y, and Z are used as accumulators for the sum of each array. These accumulators are initialized at zero before the sum loop begins. During each cycle of this loop the stored value of the variable is replaced by the stored value plus an element from the array. Thus, the six cycles sum the six elements in the array. The entire program could have been written with a single FOR-NEXT loop. Separate loops were used to facilitate clarity. The procedures in the above program closely parallel procedures in the LINEQ program.

6.3 SUBPROGRAMS

The program which is listed below calculates the best slope and intercept for a least squares line fit on a set of data points. The slope and intercept are determined by Equations (6) and (7) respectively. Some of the variables are assigned as follows:

$A = \Sigma x_i$ $\qquad F = \Sigma x_i \, \Sigma y_i$

$C = \Sigma y_i$ $\qquad G = (\Sigma x_i)^2$

$D = \Sigma x_i^2$ $\qquad H = \Sigma x_i^2 \, \Sigma y_i$

$E = \Sigma y_i x_i$

In this illustrative case, the input x-y data points fit a straight line if placed in the form

$$Y = M*(1/X) + B$$

Therefore, the computer must calculate the reciprocal of each X data point before the line-fit can be completed. The program contains a subprogram to calculate the reciprocals of X.

```
LIST
SUBPRG

10    DATA 7
20    DATA .5,4
30    DATA .333333,5
40    DATA .25,6
50    DATA .2,7
60    DATA .166667,8
70    DATA .14285,9
80    DATA .125,10
90    DIM X[20],Y[20]
100   LET A=0
110   LET C=0
120   LET D=0
130   LET E=0
140   READ N
150   FOR I=1 TO N
160   READ X[I],Y[I]
170   GOSUB 330
180   LET A=A+X[I]
190   LET C=C+Y[I]
200   LET D=D+X[I]↑2
210   LET E=E+Y[I]*X[I]
220   NEXT I
230   LET F=A*C
240   LET G=A↑2
250   LET H=C*D
260   LET M=(N*E-F)/(N*D-G)
270   LET B=(D*C-A*E)/(N*D-G)
280   PRINT "THE SLOPE IS";M
290   PRINT "THE INTERCEPT IS";B
300   STOP
310   REM THIS IS THE SUBPROGRAM TO CALCULATE THE
320   REM           RECIPROCAL OF X
330   LET X[I]=1/X[I]
340   RETURN
350   END
RUN
SUBPRG

THE SLOPE IS .999976
THE INTERCEPT IS 2.00006

DONE
```

Statements 10 to 80 are data input statements; statements 90 to 300 are the mainline program which compute the summations and calculate the slope and intercept; and statements 310 to 340 are the subprogram. To observe the use of a subprogram, refer to the FOR-NEXT loop in statement 150. The computer reads an x-y data pair in statement 160. Statement 170 directs the computer to go to the subprogram which

The Program LineQ

starts at line 310. The computer executes the subprogram, which calculates the reciprocal of the x data point, and moves to statement 340 which directs the computer to RETURN to the mainline program. The computer resumes execution of the mainline program at the statement immediately after the GO SUB statement. Each time a data pair is read, the subprogram must be called to calculate the necessary reciprocal. The computer program LINEQ uses a number of different subprograms to calculate reciprocals of X and Y and to calculate logarithms of X and Y.

6.4 USING LINEQ

The computer program LINEQ best fits experimental data with a straight line by the least squares method. Experimental x-y data points are fed into the computer program and the computer outputs the slope and intercept of the best-fit line. The computer also determines the standard deviations for the slope and intercept, it calculates a correlation coefficient and prints out a difference table. LINEQ is a frequently used program and should be a permanent resident of the computer library.

Seven different Equation Types are available as listed below:

1. Y = M*X + B
2. Y = M*(1/X) + B
3. LOG(Y) = M*X + B
4. LOG Y = M*(1/X) + B
5. (1/Y) + M*X + B
6. LOG(Y) = M*LOG(X) + B
7. Y = M*LOG(X) + B

The operator inputs only the X and Y data pairs. The computer transforms the data so that Equation Types (2) to (7) have the same final form as Equation Type (1). The user does not have to calculate quantities such as (1/X) or LOG(Y). The computer completes these calculations with appropriate subprograms.

Instructions for the use of LINEQ are outlined below. Sample computer statements are furnished to clarify the instruction.

1. Call program from computer library.
   GET-LINEQ

2. Use DATA statement to input N, the number of data pairs. Use additional DATA statements to input the values of the X-Y pairs. Use statement numbers which are less than 500. The number of data pairs must not exceed 100.

   10 DATA 7
   20 DATA .50, 4
   30 DATA .333, 5
   40 DATA .250, 6
   50 DATA .200, 7
   60 DATA .1667, 8
   70 DATA .1428, 9
   80 DATA .1250, 10

3. Type RUN. The computer will respond by asking the operator to identify the Equation Type. (A zero prints the Equation Type for operator inspection.)

4. The computer calculates values for slope intercept, standard deviations, and correlation coefficient. The computer asks the question:

   Do you want a table of deviations? (Yes, No)

   If the answer to the question is yes, then a table is printed.

```
LIST
LINEQ

500  REM FIRST DATA STATEMENT IS NUMBER OF POINTS
510  REM THEN THE POINTS ARE TYPED IN WITH THE
520  REM X-VALUES AND Y-VALUES SEPARATED BY COMMAS
530  REM
540  DIM R[100],H[100],K[100],X[100],Y[100],Z[100],Q[100]
550  READ N
560  PRINT
570  PRINT "TYPE EQUATION NUMBER OR ZERO (0.) FOR HELP";
580  INPUT H9
590  IF H9=0 THEN 650
600  IF H9<1 THEN 630
610  IF H9>7 THEN 630
620  GOTO 890
630  PRINT "NUMBER MUST BE BETWEEN 1 AND 5"
640  GOTO 560
650  PRINT
660  PRINT "THESE ARE THE LINEAR TRANFORMABLE EQUATIONS THAT"
670  PRINT "YOU HAVE TO CHOOSE FROM. PLEASE USE THE NUMBER"
680  PRINT "THAT PRECEEDS THE EQUATION."
690  PRINT
700  FOR I=1 TO 7
710  PRINT I;
720  GOSUB I OF 750,770,790,810,830,850,870
```

The Program LineQ

```
730     NEXT I
740     GOTO 560
750     PRINT "Y = M*X+B"
760     RETURN
770     PRINT "Y = M*(1/X)+B"
780     RETURN
790     PRINT "LOG(Y) = M*X+B"
800     RETURN
810     PRINT "LOG(Y) = M*(1/X)+B"
820     RETURN
830     PRINT "(1/Y) = M*X+B"
840     RETURN
850     PRINT "LOG(Y) = M*LOG(X)+B"
860     RETURN
870     PRINT "Y = M*LOG(X)+B"
880     RETURN
890     PRINT
900     PRINT " ","STANDARD DEVIATIONS FOR ";
910     GOSUB H9 OF 750,770,790,810,830,850,870
920     PRINT
930     PRINT "OF Y VALUES",
940     S=S1=S2=S3=S4=S5=S8=0
950     FOR I=1 TO N
960     READ H[I],K[I]
970     GOTO H9 OF 980,1010,1040,1070,1100,1130,1160
980     X[I]=H[I]
990     Y[I]=K[I]
1000     GOTO 1180
1010     X[I]=1/H[I]
1020     Y[I]=K[I]
1030     GOTO 1180
1040     X[I]=H[I]
1050     Y[I]=LOG(ABS(K[I]))
1060     GOTO 1180
1070     X[I]=1/H[I]
1080     Y[I]=LOG(ABS(K[I]))
1090     GOTO 1180
1100     X[I]=H[I]
1110     Y[I]=1/K[I]
1120     GOTO 1180
1130     X[I]=LOG(ABS(H[I]))
1140     Y[I]=LOG(ABS(K[I]))
1150     GOTO 1180
1160     X[I]=LOG(ABS(H[I]))
1170     Y[I]=K[I]
1180     S1=S1+X[I]
1190     S2=S2+Y[I]
1200     S3=S3+X[I]↑2
1210     S4=S4+Y[I]↑2
1220     S5=S5+X[I]*Y[I]
1230     NEXT I
1240     S6=S1↑2
1250     S7=S2↑2
1260     D1=N*S3-S6
1270     M=(N*S5-S1*S2)/D1
1280     B=(S2-M*S1)/N
1290     FOR I=1 TO N
```

```
1300    GOTO H9 OF 1310,1330,1350,1370,1430,1390,1410
1310    R[I]=M*H[I]+B
1320    GOTO 1440
1330    R[I]=M*(1/H[I])+B
1340    GOTO 1440
1350    R[I]=EXP(M*H[I]+B)
1360    GOTO 1440
1370    R[I]=EXP(M*(1/H[I])+B)
1380    GOTO 1440
1390    R[I]=EXP(B)*ABS(H[I])↑M
1400    GOTO 1440
1410    R[I]=M*LOG(ABS(H[I]))+B
1420    GOTO 1440
1430    R[I]=1/(M*H[I]+B)
1440    S=S+(K[I]-R[I])↑2
1450    NEXT I
1460    PRINT SQR(S/(N-1))
1470    T1=T2=T3=T4=T5=0
1480    FOR I=1 TO N
1490    Q[I]=M*X[I]+B
1500    S8=S8+(Y[I]-Q[I])↑2
1510    T1=T1+X[I]
1520    T2=T2+Y[I]
1530    NEXT I
1540    T1=T1/N
1550    T2=T2/N
1560    D2=(N-2)*D1
1570    M1=SQR(ABS(N*S8/D2))
1580    B1=SQR(ABS(S8*S3/D2))
1590    PRINT "OF SLOPE",M1
1600    PRINT "OF INTERCEPT",B1
1610    PRINT
1620    PRINT "VALUE OF M =",M
1630    PRINT "VALUE OF B =",B
1640    PRINT
1650    FOR I=1 TO N
1660    T3=T3+(X[I]-T1)*(Y[I]-T2)
1670    T4=T4+(X[I]-T1)↑2
1680    T5=T5+(Y[I]-T2)↑2
1690    NEXT I
1700    R=SGN(M)*T3/SQR(T4*T5)
1710    PRINT "CORRELATION COEFFICIENT =",R
1720    PRINT
1730    PRINT "DO YOU WANT A TABLE OF DEVIATIONS, (1=YES,0=NO)";
1740    INPUT H6
1750    IF H6#1 THEN 2090
1760    PRINT
1770    PRINT " ","TABLE SORTED ACCORDING TO X VALUES"
1780    PRINT
1790    PRINT "X VALUE","Y VALUE","Y PREDICTED","PERCENT DIFFERENCE"
1800    PRINT
1810    FOR I=1 TO N-1
1820    S=0
1830    FOR J=1 TO N-I
1840    IF H[J] <= H[J+1] THEN 1950
1850    S=1
1860    D=H[J+1]
```

The Program LineQ

```
1870    H[J+1]=H[J]
1880    H[J]=D
1890    D=K[J+1]
1900    K[J+1]=K[J]
1910    K[J]=D
1920    D=R[J+1]
1930    R[J+1]=R[J]
1940    R[J]=D
1950    NEXT J
1960    IF S=0 THEN 1980
1970    NEXT I
1980    FOR I=1 TO N
1990    PRINT H[I],K[I],R[I],
2000    IF K[I]=0 THEN 2070
2010    P=100*(K[I]-R[I])/K[I]
2020    IF P<0 THEN 2050
2030    PRINT " ",P
2040    GOTO 2080
2050    PRINT P
2060    GOTO 2080
2070    PRINT
2080    NEXT I
2090    END
```

6.5 SAMPLE CALCULATIONS WITH LINEQ

Fit the x-y data points as listed below with linear Equation Type 1. Study the data and decide on the acceptability of the fit. Fit the data points with linear Equation Type 2. Study the data and decide on the acceptability of the fit.

x	y
0.5	4
0.333	5
0.250	6
0.200	7
0.1667	8
0.1428	9
0.1250	10

The computer results of the Equation Type 1 line fit are given below:

GET-LINEQ

```
10 DATA 7
20 DATA .500, 4
30 DATA .333, 5
40 DATA .250, 6
50 DATA .200, 7
60 DATA .167, 8
70 DATA .143, 9
80 DATA .125, 10
RUN
LINEQ
```

TYPE EQUATION NUMBER OR ZERO (0.) FOR HELP? 1

$$\text{STANDARD DEVIATIONS FOR Y} = M*X+B$$

OF Y VALUES	.831037
OF SLOPE	2.80002
OF INTERCEPT	.768534

VALUE OF M = -15.0229
VALUE OF B = 10.687

CORRELATION COEFFICIENT = .923044

DO YOU WANT A TABLE OF DEVIATIONS, (1=YES,0=NO)? 1

TABLE SORTED ACCORDING TO X VALUES

X VALUE	Y VALUE	Y PREDICTED	PERCENT DIFFERENCE	
.125	10	8.80918		11.9081
.143	9	8.53877		5.12473
.167	8	8.17822	-2.22781	
.2	7	7.68247	-9.74956	
.25	6	6.93132	-15.5221	
.333	5	5.68443	-13.6885	
.5	4	3.1756		20.6099

DONE

The best possible Type 1 line through these experimental points has a slope of -15.023 ± 2.8 and an intercept of $10.687 \pm .769$. The poor correlation coefficient and the large differences in the table of deviations lead to the conclusion that the data does not fit this type of straight line.

The results of an Equation Type 2 line fit are much improved as indicated by the following computer output:

RUN
LINEQ

TYPE EQUATION NUMBER OR ZERO (0.) FOR HELP? 2

$$\text{STANDARD DEVIATIONS FOR Y} = M*(1/X)+B$$

OF Y VALUES	4.63191E-03
OF SLOPE	9.59992E-04
OF INTERCEPT	5.16687E-03

VALUE OF M = 1.00114
VALUE OF B = 1.99658

CORRELATION COEFFICIENT = .999998

DO YOU WANT A TABLE OF DEVIATIONS, (1=YES,0=NO)? 1

The Program LineQ

TABLE SORTED ACCORDING TO X VALUES

X VALUE	Y VALUE	Y PREDICTED	PERCENT DIFFERENCE	
.125	10	10.0057	-5.70679E-02	
.143	9	8.99756		2.70632E-02
.167	8	7.99144		.107062
.2	7	7.00228	-3.26293E-02	
.25	6	6.00114	-1.90417E-02	
.333	5	5.00301	-6.01578E-02	
.5	4	3.99886		2.85029E-02

DONE

The best possible Type 2 line fit has a slope of 1.0011 ± .0009 and an intercept of 1.9966 ± .0052. The correlation coefficient is nearly one. The table of deviations indicates an excellent fit on every data point. Clearly, then, the data fits a Type 2 straight line.

6.6 EXERCISES

1. The vapor pressures of a gas are described by the equation

$$\log P = \frac{A}{T} + B$$

where P is vapor pressure and T is absolute temperature. Given a list of vapor pressures for ethanol, calculate the A and B values.

T(°K)	P(mm of Hg)
283	23.5
293	44.0
303	78.9
313	135.2
323	222.3
333	352.7
343	542.4

2. A gas at moderate pressures follows the equation of state

$$PV = bP + RT$$

where P is pressure, V is volume, and T is absolute temperature. Given the following data for a gas at 298.15°K, determine the value of b and the value of R in the above equation.

P (atm)	PV (1. atm.)
0	24.454
1	24.421
2	24.388
3	24.355
4	24.322
5	24.289
6	24.256

3. Given the following x-y data pairs. Attempt to fit the data by a straight line equation. Which Equation Type best describes this data?

X	Y
0	.08
11.2	.0788
22.7	.0776
32.5	.0766
41.5	.0757
47.6	.0751
59.0	.0740
66.4	.0733
75.0	.0725
93.8	.0708

CHAPTER 7

The Program PoleQ

7.1 DESCRIPTION OF POLEQ

The program POLEQ fits a set of plotted x and y data points with a polynomial by least squares method. It determines the coefficients of a polynomial of the form

$$y = a_0 + a_1 x + a_2 x^2 + a_3 x^3 + \cdots \qquad (1)$$

POLEQ will easily fit data points with polynomials up to approximately eighth order. It will fit certain sets of data with polynomials up to twenty-fifth order. It also aids the operator in judging the accuracy of the curve fit by calculating a standard deviation and by listing a difference table. POLEQ is a frequently used program and should be a permanent resident in a computer library.

The method of least squares will be illustrated by using a third degree polynomial. Data points x_1, y_1; x_2, y_2; x_3, y_3; $\cdots$; x_n, y_n are fit with an equation

$$y = a_0 + a_1 x + a_2 x^2 + a_3 x^3 \qquad (2)$$

Any number of different third degree polynomials might be considered. The coefficients a_0, a_1, a_2, and a_3 for the best polynomial, however, are to be determined. As with linear least squares, the residues between experimental and calculated data points are squared and summed. Therefore,

$$S = \sum_{i=1}^{n} [y_i - a_0 - a_1 x_i - a_2 x_i^2 - a_3 x_i^3]^2 \qquad (3)$$

where the summation is completed over n data points, where each y_i is an experimental data point, and where the remaining terms calculate each y value from Equation (2). The best polynomial produces a minimum value of S.

Since S if a function of a_0, a_1, a_2, and a_3, the following equations must hold at the minimum:

$$\frac{\delta S}{\delta a_0} = 0, \quad \frac{\delta S}{\delta a_1} = 0, \quad \frac{\delta S}{\delta a_2} = 0, \quad \frac{\delta S}{\delta a_3} = 0$$

These partial derivatives may be rewritten as:

$$a_0 \Sigma x_i^0 + a_1 \Sigma x_i + a_2 \Sigma x_i^2 + a_3 \Sigma x_i^3 = \Sigma y_i$$

$$a_0 \Sigma x_i + a_1 \Sigma x_i^2 + a_2 \Sigma x_i^3 + a_3 \Sigma x_i^4 = \Sigma x_i y_i$$

$$a_0 \Sigma x_i^2 + a_1 \Sigma x_i^3 + a_2 \Sigma x_i^4 + a_3 \Sigma x_i^5 = \Sigma x_i^2 y_i$$

$$a_0 \Sigma x_i^3 + a_1 \Sigma x_i^4 + a_2 \Sigma x_i^5 + a_3 \Sigma x_i^6 = \Sigma x_i^3 y_i$$

These four equations contain summations of the known experimental data points x_i and y_i and they contain four unknown coefficients a_0, a_1, a_2, and a_3. The simultaneous solution of four equations produces the values of the coefficients for a third degree polynomial where the sum of the squared deviations is a minimum. The program POLEQ determines the coefficients by a technique called Gaussian elimination. Any textbook in numerical analysis can provide further details on Gaussian elimination.

7.2 USING POLEQ

The operator inputs up to a hundred x-y data pairs and enters the command RUN. The operator then selects the order of the polynomial. He may select first order which is simply a line-fit; he may select second order where the data is fit by a quadratic equation; he may select third order where the highest term in the polynomial is a cubic; or he may select higher orders if necessary. In most cases, it

The Program PoleQ

is advisable to RUN several different polynomial fits where different orders are selected, and to choose the fit which yields the best results. The standard deviations and the difference tables assist in judging the most acceptable fit.

The detailed instructions for the use of POLEQ are listed below. Sample computer statements are included to clarify the instructions.

1. Call program from library.

    GET-POLEQ

2. A first DATA statement is used to inform the computer of the number of data pairs to be used. Additional DATA statements are used to input the values of the data pairs. Statement numbers for DATA statements must be less than 500.

    ```
    10 DATA 7
    20 DATA 0,1.000
    30 DATA 50, .9082
    40 DATA 100, .8352
    50 DATA 150, .7823
    60 DATA 200, .7508
    70 DATA 250, .7431
    80 DATA 300, .7621
    ```

3. Type RUN. The computer responds with the following question:

    Type highest power of X in polynomial that you want fit through these points.

    The operator answers the question by typing 1, 2, 3, $\cdots$ to designate first, second, third, $\cdots$ order.

4. The computer calculates the coefficients and the standard deviation for the fit. The following question then appears:

    Do you want a table of deviations?

    A yes answer provides the table. The table may also be used to verify that input data is correct.

```
LIST
POLEQ

500    DIM X[100],Y[100],A[625],B[25],F[100]
510    READ N
520    S1=0
530    FOR I=1 TO N
540    READ X[I],Y[I]
```

```
550    S1=S1+Y[I]
560    NEXT I
570    PRINT
580    PRINT "TYPE HIGHEST POWER OF X IN POLYNOMIAL"
590    PRINT "THAT YOU WANT FIT THROUGH THESE POINTS";
600    INPUT M
610    IF M >= N THEN 650
620    IF M <= 0 THEN 650
630    IF M#INT(M) THEN 650
640    GOTO 670
650    PRINT "THIS MUST BE AN INTEGER > ZERO AND LESS THAN";N
660    GOTO 570
670    PRINT
680    PRINT "STANDARD DEVIATION OF ACTUAL AND PREDICTED POINTS =";
690    FOR I=1 TO M+1
700    FOR J=1 TO M+1
710    IF (J=1) AND (I=1) THEN 770
720    S=0
730    FOR L=1 TO N
740    S=S+X[L]↑(I+J-2)
750    NEXT L
760    A[I+(M+1)*(J-1)]=S
770    NEXT J
780    NEXT I
790    A[1]=N
800    FOR I=2 TO M+1
810    S=0
820    FOR J=1 TO N
830    S=S+Y[J]*X[J]↑(I-1)
840    NEXT J
850    B[I]=S
860    NEXT I
870    B[1]=S1
880    REM ROUTINE TO LINEAR EQUATIONS
890    REM
900    M1=M+1
910    N0=M1
920    REM FORWARD SOLUTION
930    T1=0
940    K1=0
950    J1=-M1
960    FOR J0=1 TO M1
970    J2=J0+1
980    J1=J1+M1+1
990    B1=0
1000   I1=J1-J0
1010   FOR I0=J0 TO M1
1020   REM SEARCH FOR MAX COEFF IN COLUMN
1030   I2=I1+I0
1040   IF ABS(B1) >= ABS(A[I2]) THEN 1070
1050   B1=A[I2]
1060   I3=I0
1070   NEXT I0
1080   REM TEST FOR SINGULAR MATRIX
1090   IF ABS(B1)>T1 THEN 1130
1100   K1=1
1110   PRINT "SINGULAR MATRIX / CHECK DATA"
```

```
1120    STOP
1130    L1=J0+M1*(J0-2)
1140    I1=I3-J0
1150    FOR K0=J0 TO M1
1160    L1=L1+M1
1170    L2=L1+I1
1180    S1=A[L1]
1190    A[L1]=A[L2]
1200    A[L2]=S1
1210    A[L1]=A[L1]/B1
1220    NEXT K0
1230    S1=B[I3]
1240    B[I3]=B[J0]
1250    B[J0]=S1/B1
1260    REM
1270    IF J0=N0 THEN 1410
1280    I4=N0*(J0-1)
1290    FOR I6=J2 TO N0
1300    I5=I4+I6
1310    I1=J0-I6
1320    FOR J4=J2 TO M1
1330    I7=M1*(J4-1)+I6
1340    J3=I7+I1
1350    A[I7]=A[I7]-(A[I5]*A[J3])
1360    NEXT J4
1370    B[I6]=B[I6]-(B[J0]*A[I5])
1380    NEXT I6
1390    NEXT J0
1400    REM BACK SOLUTION
1410    N1=M1-1
1420    I1=M1*M1
1430    FOR J0=1 TO N1
1440    Z1=I1-J0
1450    Z2=N0-J0
1460    Z3=N0
1470    FOR K0=1 TO J0
1480    B[Z2]=B[Z2]-A[Z1]*B[Z3]
1490    Z1=Z1-M1
1500    Z3=Z3-1
1510    NEXT K0
1520    NEXT J0
1530    REM ***** END OF LINEAR SOLVER ROUTINE
1540    S2=0
1550    FOR I=1 TO N
1560    S=B[1]
1570    FOR J=1 TO M
1580    S=S+B[J+1]*X[I]↑J
1590    NEXT J
1600    E[I]=S
1610    S2=S2+(Y[I]-S)↑2
1620    NEXT I
1630    S2=SQR(S2/(N-1))
1640    PRINT S2
1650    PRINT
1660    PRINT
1670    PRINT "COEFFICIENTS OF POLYNOMIAL"
1680    FOR I=1 TO M
```

```
1690   PRINT
1700   PRINT B[M-I+2];"* X !";M-I+1
1710   NEXT I
1720   PRINT
1730   PRINT "PLUS CONSTANT";B[1]
1740   PRINT
1750   PRINT
1760   PRINT "DO YOU WANT A TABLE OF DEVIATIONS (0=NO,1=YES)";
1770   INPUT T
1780   IF T=0 THEN 2120
1790   PRINT
1800   PRINT " ","TABLE SORTED ACCORDING TO X VALUES"
1810   PRINT
1820   PRINT "X-VALUE","Y-VALUE","Y FIT","PERCENT DIFFERENCE"
1830   PRINT
1840   FOR I=1 TO N-1
1850   S=0
1860   FOR J=1 TO N-I
1870   IF X[J]<X[J+1] THEN 1980
1880   S=1
1890   D=X[J+1]
1900   X[J+1]=X[J]
1910   X[J]=D
1920   D=Y[J+1]
1930   Y[J+1]=Y[J]
1940   Y[J]=D
1950   D=E[J+1]
1960   E[J+1]=E[J]
1970   E[J]=D
1980   NEXT J
1990   IF S=0 THEN 2010
2000   NEXT I
2010   FOR I=1 TO N
2020   PRINT X[I],Y[I],E[I],
2030   IF Y[I]=0 THEN 2100
2040   P=100*(Y[I]-E[I])/Y[I]
2050   IF P>0 THEN 2080
2060   PRINT P
2070   GOTO 2110
2080   PRINT " ",P
2090   GOTO 2110
2100   PRINT
2110   NEXT I
2120   END
```

7.3 SAMPLE RUN WITH POLEQ

The compressibilities of a gas are given at various pressures. Describe the data in polynomial form such as

$$Z = A + BP + CP^2 + DP^3$$

where Z is a compressibility and P is pressure in atmospheres.

The Program PoleQ

P (atm)	Z
0	1.00000
50	.90817
100	.835226
150	.782257
200	.750816
250	.743141
300	.762106
350	.811227
400	.894656
450	1.01719

```
GET-POLEQ

10   DATA 10
20   DATA 0,1
30   DATA 50,.90817
40   DATA 100,.835226
50   DATA 150,.782257
60   DATA 200,.750816
70   DATA 250,.743141
80   DATA 300,.762106
90   DATA 350,.811227
100  DATA 400,.894656
110  DATA 450,1.01719
RUN
POLEQ


TYPE HIGHEST POWER OF X IN POLYNOMIAL
THAT YOU WANT FIT THROUGH THESE POINTS? 3

STANDARD DEVIATION OF ACTUAL AND PREDICTED POINTS = 1.13361E-03


COEFFICIENTS OF POLYNOMIAL

 3.83638E-09    * X ↑ 3

 2.64288E-06    * X ↑ 2

-1.92797E-03    * X ↑ 1

PLUS CONSTANT  .998859


DO YOU WANT A TABLE OF DEVIATIONS (0=NO,1=YES)? 1
```

TABLE SORTED ACCORDING TO X VALUES

X-VALUE	Y-VALUE	Y FIT	PERCENT DIFFERENCE	
0	1	.998859		.114071
50	.90817	.909547	-.151675	
100	.835226	.836327	-.13188	
150	.782257	.782076		2.31025E-02
200	.750816	.749671		.15247
250	.743141	.74199		.154895
300	.762106	.761909		2.57938E-02
350	.811227	.812307	-.133107	
400	.894656	.89606	-.15691	
450	1.01719	1.01605		.11253

DONE

The coefficients are listed by the computer. The standard deviation between calculated and experimental y-values is .00011. The fit is highly acceptable, in that the difference table shows a close fit at every data point.

7.4 EXERCISES

1. Calculated heat capacities for gaseous benzene are listed below. Fit the data with a polynomial of the form

$$C_p = A + BT + CT^2 + DT^3$$

where C_p is heat capacity and T is absolute temperature.

T(°K)	C_p (cal/mole)
400	26.74
600	37.65
800	44.91
1000	49.98
1200	53.67
1400	56.40
1600	58.45
1800	60.02
2000	61.23

2. Express the experimental data which follow in polynomial form where

$$PV = A + BP + CP^2$$

PV(1 atm)	P (atm)
0	30.62
50	26.87
100	22.93
150	19.54
200	18.13
250	18.60
300	19.81

3. Complete several different polynomial fits on the x-y data points which follow. Which polynomial order sufficiently describes the data?

X	Y
1	8.2
2	11.8
3	17.8
4	26.2
5	37.0
6	50.2
7	65.8
8	83.8

CHAPTER 8

The Program Roots

8.1 DESCRIPTION OF ROOTS

The computer program ROOTS determines the roots of a polynomial in the form
$$y = a_0 + a_1 x + a_2 x^2 + a_3 x^3 + \cdots \tag{1}$$
by Newton's method of successive approximations. Newton's method will be illustrated for the case where a real root is being determined. Consider the plot of an arbitrary function, $f(x)$, as illustrated in Figure 8.1. The one real root of this function is illustrated. It occurs at the point where the curve crosses the x axis; at this value of x the function $f(x)$ has a zero value.

The computer program will need to determine the value of x for which the function has a zero value. The computer is programmed to guess an initial value of x as being the root. Let this initial guess for the value of the root be labelled as an x_0. Clearly the initial guess has little chance of being the actual root. The program, however, will improve on this initial guess by using one cycle of Newton's method. This produces a second improved "guess" for the value of the root and is labelled x_1. The entire process is repeated again in a second cycle of Newton's method to obtain a third "guess" at the value of the root. This cyclic or reiterative method is continued until the "guessed" value of x_0 is really the desired root of the polynomial.

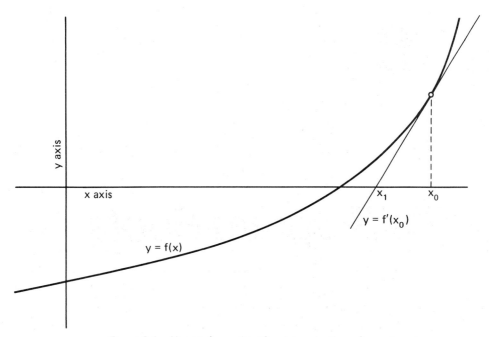

Figure 8.1 Newton's method for determination of a real root

To illustrate one cycle of Newton's method consider again the initial guess of x_0. The function is evaluated at this point and labelled $f(x_0)$. The value of $f(x_0)$ is clearly not zero. A tangent is drawn to the function at $x = x_0$. This straight line now approximates the curvature of the function. The place where the straight line crosses the x-axis will be the improved guess of the root and is temporarily labelled x_1. The mathematical formula to determine x_1 is developed as follows:

$$\text{slope} = f'(x_0) = \frac{f(x_0) - f(x_1)}{(x_0 - x_1)} \qquad (2)$$

The straight line is the tangent of the function at x_0 and slope is therefore also the first derivative of the function at x_0. The slope is also the $\Delta y/\Delta x$ and can therefore be defined in terms of the two points $(x_0, f(x_0))$ and $(x_1, f(x_1))$. The value of $f(x_1)$ is always zero since it appears on the x axis, therefore Equation (2) may be rewritten as

$$(x_0 - x_1) = \frac{f(x_0)}{f'(x_0)} \qquad (3)$$

The improved guess of the roots is then given by

$$x_1 = x_0 - \frac{f(x_0)}{f'(x_0)} \quad (4)$$

This value of x_1 is now considered to be the new guess of the root and is labelled x_0. An improved value for x_1 is sought on a second iteration. This simple procedure is extended into the imaginary plane so as to solve for all real and complex roots of a polynomial.

8.2 USING ROOTS

ROOTS determines all of the roots of most polynomials which do not exceed fifteenth order. It is capable of handling even higher orders if the polynomials are well behaved. ROOTS has been found to be extremely versatile and capable of solving all polynomials which are proposed in this textbook. Certain polynomials may exist, however, which may go beyond the limits of the ROOTS program.

If a chemist is finding the roots of a fourth order polynomial, the equation has the form

$$y = a_0 + a_1 x + a_2 x^2 + a_3 x^3 + a_4 x^4 \quad (5)$$

The program operator will describe the polynomial for the ROOTS program by specifying that the polynomial is fourth order and by inputting the five coefficients. The computer program automatically calculates the four roots of this polynomial. The real roots are labelled and printed followed by the complex roots.

Instructions for the use of ROOTS are outlined below. The sample computer statements describe a fourth degree polynomial where the coefficients are given as follows:

$a_0 = -21.345$

$a_1 = -2.1$

$a_2 = 0.34$

$a_3 = 3.0$

$a_4 = 1.0$

1. Call program from computer library.

    GET-ROOTS

The Program Roots

2. Type RUN and describe a polynomial by answering the following questions. Sample teletype inputs are enclosed.

   a. The highest degree of the polynomial?   4
   b. The constant term (a_0)?   -21.345
   c. The first degree coefficient (a_1)?   -2.1
   d. The second degree coefficient (a_2)?   .34
   e. The third degree coefficient (a_3)?   3.0
   f. The fourth degree coefficient (a_4)?   1.0

3. The computer calculates and outputs the roots. In this example the calculated real roots are -3.29863 and 1.71856. The complex roots are expressed as:

   .709966 - 1.80589 $\sqrt{-1}$

   .709966 + 1.80589 $\sqrt{-1}$

```
LIST
ROOTS

1    PRINT "WHAT IS THE HIGHEST DEGREE OF X IN THE POLYNOMIAL";
2    INPUT M
3    PRINT
4    IF INT(M)#M THEN 8
5    IF M<1 THEN 8
6    IF M>31 THEN 8
7    GOTO 10
8    PRINT "THIS MUST BE AN INTEGER > THAN 0 AND < 31"
9    GOTO 1
10   PRINT "WHAT IS THE CONSTANT TERM";
11   INPUT X[1]
12   F[1]=X[1]
13   PRINT
14   PRINT "COEFFICIENTS"
15   FOR I=2 TO M+1
16   PRINT
17   PRINT "X ↑";I-1;
18   INPUT X[I]
19   F[I]=X[I]
20   NEXT I
21   FOR I=1 TO M
22   R[I]=0
23   Z[I]=0
24   NEXT I
25   PRINT
26   FOR I=1 TO 60
27   PRINT "-";
28   NEXT I
29   DIM X[31],C[31],R[30],Z[30],F[31],G[3]
30   I1=0
31   N1=M
```

```
32   IF X[N1+1]=0 THEN 34
33   GOTO 36
34   PRINT "THE COEFFICIENT ON THE HIGHEST DEGREE WAS ZERO."
35   STOP
36   N2=N1
37   N3=N1+1
38   N4=1
39   K1=N1+1
40   FOR L=1 TO K1
41   M1=K1-L+1
42   C[M1]=X[L]
43   NEXT L
44   X0=5.00101E-03
45   Y0=.010001
46   I2=0
47   X1=X0
48   X0=-10*Y0
49   Y0=-10*X1
50   X1=X0
51   Y1=Y0
52   I2=I2+1
53   GOTO 57
54   I1=1
55   X2=X1
56   Y2=Y1
57   I3=0
58   U1=0
59   U2=0
60   V=0
61   Y3=0
62   X3=1
63   U=C[N1+1]
64   IF U=0 THEN 115
65   FOR I=1 TO N1
66   L=N1-I+1
67   X4=X1*X3-Y1*Y3
68   Y4=X1*Y3+Y1*X3
69   U=U+C[L]*X4
70   V=V+C[L]*Y4
71   U1=U1+I*X3*C[L]
72   U2=U2-I*Y3*C[L]
73   X3=X4
74   Y3=Y4
75   NEXT I
76   S=U1*U1+U2*U2
77   IF S=0 THEN 106
78   D1=(V*U2-U*U1)/S
79   X1=X1+D1
80   D2=-(U*U2+V*U1)/S
81   Y1=Y1+D2
82   IF ABS(X1)<ABS(Y1) THEN 85
83   Q9=ABS(X1)
84   GOTO 86
85   Q9=ABS(Y1)
86   IF Q9>.1 THEN 88
87   Q9=.1
```

```
88    IF ABS(D1)+ABS(D2)<Q9*.00001 THEN 95
89    I3=I3+1
90    IF I3<500 THEN 58
91    IF I1#0 THEN 95
92    IF I2<5 THEN 47
93    PRINT "NO ROOTS FOUND AFTER 500 TRYS ON 5 STARTING VALUES."
94    STOP
95    FOR L=1 TO N3
96    M1=K1-L+1
97    T=X[M1]
98    X[M1]=C[L]
99    C[L]=T
100   NEXT L
101   I4=N1
102   N1=N2
103   N2=I4
104   IF I1=0 THEN 54
105   GOTO 109
106   IF I1=0 THEN 47
107   X1=X2
108   Y1=Y2
109   I1=0
110   IF ABS(Y1/X1)<.0001 THEN 118
111   A=X1+X1
112   S=X1*X1+Y1*Y1
113   N1=N1-2
114   GOTO 122
115   X1=0
116   N2=N2-1
117   N3=N3-1
118   Y1=0
119   S=0
120   A=X1
121   N1=N1-1
122   C[2]=C[2]+A*C[1]
123   FOR L=2 TO N1
124   C[L+1]=C[L+1]+A*C[L]-S*C[L-1]
125   NEXT L
126   Z[N4]=Y1
127   R[N4]=X1
128   N4=N4+1
129   IF S=0 THEN 133
130   Y1=-Y1
131   S=0
132   GOTO 126
133   IF N1>0 THEN 44
134   FOR I=1 TO M
135   IF Z[I]#0 THEN 154
136   IF M<I+1 THEN 155
137   IF Z[I+1]#0 THEN 154
138   IF ABS(R[I]-R[I+1])>ABS(R[I]+R[I+1])*.01 THEN 154
139   G[3]=(R[I]+R[I+1])*.5
140   FOR J=0 TO 2
141   C[M+1]=F[M+1]
142   IF J=2 THEN 144
143   G[J+1]=R[I+J]
```

```
144     FOR L=1 TO M
145     J5=M-L+1
146     C[J5]=F[J5]+G[J+1]*C[J5+1]
147     NEXT L
148     X[J+1]=C[1]
149     NEXT J
150     IF ABS(X[1]) <= ABS(X[3]) THEN 154
151     IF ABS(X[2]) <= ABS(X[3]) THEN 154
152     R[I]=(R[I]+R[I+1])*.5
153     R[I+1]=R[I]
154     NEXT I
155     PRINT
156     PRINT
157     S=0
158     FOR I=1 TO M
159     IF Z[I]#0 THEN 161
160     S=S+1
161     NEXT I
162     IF S>0 THEN 166
163     PRINT "THERE ARE NO REAL ROOTS OF THIS POLYNOMIAL"
164     PRINT
165     GOTO 174
166     PRINT "THESE ARE THE REAL ROOTS OF THE POLYNOMIAL"
167     PRINT
168     FOR I=1 TO M
169     IF Z[I]#0 THEN 172
170     PRINT
171     PRINT R[I]
172     NEXT I
173     PRINT
174     IF S<M THEN 178
175     PRINT "THERE ARE NO COMPLEX ROOTS OF THE POLYNOMIAL"
176     PRINT
177     GOTO 185
178     PRINT "THESE ARE THE COMPLEX ROOTS OF THE POLYNOMIAL"
179     PRINT
180     FOR I=1 TO M
181     IF Z[I]=0 THEN 184
182     PRINT
183     PRINT R[I],"PLUS",Z[I],"TIMES SQR(-1)"
184     NEXT I
185     END
```

8.3 SAMPLE CALCULATION

Calculate the volume of 200 grams of N_2 gas at $300°K$ and 100 atmospheres pressure. Use van der Waals equation which can be written in the form

$$PV^3 - (nbP + nRT)V^2 + an^2V - an^3b = 0 \qquad (6)$$

The Program Roots

where P is pressure in atmospheres, V is volume in liters, T is absolute temperature and where a, b, and R are constants which are defined as follows for N_2:

$a = .81$
$b = .030$
$R = .064$

Equation (6) can be written as

$$AV^3 + BV^2 + CV + D = 0 \qquad (7)$$

where $A = P = 100$
$B = -nbP - nRT = -115.668$
$C = an^2 = 41.29$
$D = -an^3b = -8.845$

The program roots is applied to Equation (7).

```
RUN
ROOTS

WHAT IS THE HIGHEST DEGREE OF X IN THE POLYNOMIAL? 3

WHAT IS THE CONSTANT TERM? -8.845

COEFFICIENTS

X ↑ 1      ? 41.29

X ↑ 2      ? -115.668

X ↑ 3      ? 100
-----------------------------------------------------------
THESE ARE THE REAL ROOTS OF THE POLYNOMIAL

 .769462

THESE ARE THE COMPLEX ROOTS OF THE POLYNOMIAL

 .193609        PLUS      -.278327       TIMES SQR(-1)

 .193609        PLUS       .278327       TIMES SQR(-1)

DONE
```

The polynomial has three roots. The complex roots can be neglected in this real physical problem; therefore, the volume of the gas must be 0.769 lliters.

8.4 EXERCISES

1. Given the polynomial

$$y = x^4 - 3x^3 - 49x^2 - 57x + 108 \qquad (6)$$

   Find all of the roots of this polynomial.

2. The compressibility of an ethane gas is given by

$$Z = 1 + AP + BP^2 + CP^3 \qquad (7)$$

   where Z is compressibility, P is pressure in atmospheres, and where the coefficients are defined as follows:

   $A = -7.23 \times 10^{-3}$
   $B = 15.0 \times 10^{-6}$
   $C = 4.02 \times 10^{-9}$

   a. At what pressure does the compressibility have a minimum value? Remember that for a minimum the first derivative has a zero value. Differentiate Equation (7), set result equal to zero and find real roots for desired pressure.

   b. Find the two pressures where the gas is ideal (Z has a value of one) Solve by finding the values of P where

$$AP + BP^2 + CP^3 = 0 \qquad (8)$$

3. Acetic acid dissociates according to the equation

$$HAc \rightarrow H^+ + Ac^-$$

   The equilibrium constant is defined as

$$K_a = \frac{[H^+][Ac^-]}{[HAc]} = \frac{c\alpha^2}{1 - \alpha}$$

   where c is the concentration in moles per liter and α is the percent dissociation. The K_a for acetic acid is 1.8×10^{-5}. Calculate the percentage dissociation of an acetic acid solution which is 10^{-4} molar.

CHAPTER 9

Gas Calculations

9.1 GAS DENSITIES FROM IDEAL GAS LAW

The ideal gas law in its most familiar form is written as

$$PV = nRT \tag{1}$$

where P is pressure, V is volume, n is moles, T is absolute temperature and R is the ideal gas constant. Equation (1) may be rewritten as

$$PV = \frac{Wt}{M} RT \tag{2}$$

where M is the gram molecular weight and Wt is the weight of the sample. Rearrangement yields

$$P = \frac{Wt}{V} \frac{RT}{M} = D \frac{RT}{M} \tag{3}$$

where D is the density of the gas. Finally,

$$D = \frac{PM}{RT} \tag{4}$$

When pressures are expressed in atmospheres, M in grams per mole and T in degrees Kelvin; R has a value of .08206 liter atmospheres per mole per degree. Resulting densities are expressed in grams per liter.

1. Write a simple computer program to calculate the densities of an ideal gas at pressures of 5, 10, 15, 20, and 25 atmospheres and at temperatures of 300, 400, 500, 600, and 700 degrees Kelvin. Use a FOR-NEXT LOOP to determine densities for each of the listed temperatures. Use a second FOR-NEXT LOOP nested within the first loop to determine densities for each of the listed pressures. Use the program to calculate the densities of H_2 gas and to calculate the densities of methane gas.

9.2 VAN DER WAALS EQUATION OF STATE

A familiar form of van der Waals equation of state for real gases is

$$\left(P + \frac{an^2}{V^2} \right)(V - nb) = nRT \tag{5}$$

The equation is easily rewritten to solve for pressure.

$$P = \frac{nRT}{(V - nb)} - \frac{an^2}{V^2} \tag{6}$$

A second rearrangement according to the following steps illustrates that solving for a volume within van der Waals equation requires the roots of a cubic equation:

$$\left(\frac{PV^2 + an^2}{V^2} \right)(V-nb) = nRT \tag{7}$$

$$(PV^2 + an^2)(V-nb) = nRTV^2 \tag{8}$$

$$PV^3 - nbPV^2 + an^2V - an^3b = nRTV^2 \tag{9}$$

$$PV^3 - (nRT + nbP)V^2 + an^2V - an^3b = 0 \tag{10}$$

The polynomial form of the last equation is most evident if written as follows:

$$AV^3 + BV^2 + CV + D = 0 \tag{11}$$

where
  $A = P$
  $B = -nRT - nbP$
  $C = an^2$
  $D = -an^3b$

Gas Calculations

TABLE 9.1
Van der Waals constants for some common gases.
The value of R is .08206 liter atmospheres per mole per degree.

Substance	$a(1^2 \text{ atm mole}^{-2})$	$b = \text{liters mole}^{-1}$
He	.0341	.0237
H_2	.2444	.0266
N_2	1.390	.0391
O_2	1.360	.0318
CH_4	2.253	.0428
C_2H_4	5.489	.0638
CO	1.485	.0399
CO_2	3.592	.0427
Ne	.2107	.0171
Ar	1.345	.03219

1. Calculate the pressure of a mole of N_2 gas when it is confined to 200 milliliters of volume at 273.15°K. Compare to the ideal gas pressure. Repeat the calculation for several other gases such as carbon dioxide, oxygen, and hydrogen. [Prepare your own program.]

2. Calculate the volume of a mole of nitrogen gas which is confined at 200 atmospheres and 500°K. Compare to the ideal gas volume. Repeat the calculations for several other gases such as carbon dioxide, oxygen, and hydrogen. [Use ROOTS to solve van der Waals equation for volume.]

9.3 VIRIAL EQUATION FOR REAL GASES

An improved equation of state for a mole of real gas has the form

$$PV = RT + AP + BP^2 + CP^3 + DP^4 \qquad (12)$$

The coefficients, called virial coefficients, are temperature dependent. When solving this equation for volume the equation is arranged as follows:

$$V = \frac{RT}{P} + A + BP + CP^2 + DP^3 \qquad (13)$$

When solving the virial equation for pressure, the polynomial is rearranged so that

$$DP^4 + CP^3 + BP^2 + (A-V)P + RT = 0 \qquad (14)$$

The positive real root of the above equation corresponds to the solution of the problem.

Maron & Turnbull have investigated the temperature dependence of the virial coefficients for nitrogen gas. They submit the following equations to describe the coefficients as a function of the absolute temperature.

$$A = a_1 + \frac{a_2}{T} + \frac{a_3}{T^3} \tag{15}$$

$$B = \frac{b_1}{T^2} + \frac{b_2}{T^4} + \frac{b_3}{T^6} \tag{16}$$

$$C = \frac{c_1}{T^2} + \frac{c_2}{T^4} + \frac{c_3}{T^6} \tag{17}$$

$$D = \frac{d_1}{T^2} + \frac{d_2}{T^4} + \frac{d_3}{T^6} \tag{18}$$

TABLE 9.2
Empirical Constants for the Determination of Virial Coefficients of Nitrogen Gas

$a_1 = 3.835 \times 10^{-2}$	$c_1 = -6.660 \times 10^{-3}$
$a_2 = -10.07$	$c_2 = 3.920 \times 10^{2}$
$a_3 = -2.449 \times 10^{5}$	$c_3 = -2.089 \times 10^{7}$
$b_1 = 6.594$	$d_1 = 2.703 \times 10^{-6}$
$b_2 = -1.217 \times 10^{5}$	$d_2 = -0.2829$
$b_3 = 8.217 \times 10^{9}$	$d_3 = 1.314 \times 10^{4}$

When the constants from Table 9.2 are substituted into Equations (15) - (18), the coefficients for nitrogen at 273.15°K assume the following values:

RT = 22.415 liter atmospheres per mole

$A = -1.053 \times 10^{-2}$

$B = 8.630 \times 10^{-5}$

$C = -6.914 \times 10^{-8}$

$D = 1.704 \times 10^{-11}$

Similar coefficients may be calculated at any other temperature.

Gas Calculations

1. Calculate the pressure of a mole of N_2 gas when it is confined to 200 milliliters of volume at 273.15°K. Compare to the ideal gas pressure. [Use ROOTS to solve the virial equation as described in Equation (14).]

2. Calculate the volume of a mole of nitrogen gas which is confined at 200 atmospheres and 500°K. Compare to the ideal gas volume. [Prepare your own program to evaluate the virial coefficients at 500°K.]

3. Prepare a pressure-volume plot for nitrogen at 273.15°K. Prepare a similar plot for nitrogen gas at 500°K. [Use PLOT to prepare your graphs. Use pressure as the independent variable in Equation (13).]

9.4 BEATTIE-BRIDGEMAN FORM OF VIRIAL EQUATION

One of the most successful equations of state for a gas has the form

$$P = \frac{\alpha n}{V} + \frac{\beta n^2}{V^2} + \frac{\gamma n^3}{V^3} + \frac{\delta n^4}{V^4} \tag{19}$$

The coefficients of α, β, γ, and δ depend on the nature of the gas and on the temperature. The coefficients may be defined as follows:

$$\alpha = RT \tag{20}$$

$$\beta = -A_o + B_o RT - \frac{cR}{T^2} \tag{21}$$

$$\gamma = aA_o - bB_o RT - \frac{cB_o R}{T^2} \tag{22}$$

$$\delta = \frac{bcB_o R}{T^2} \tag{23}$$

The coefficients for various gases are defined by using the Beattie and Bridgeman constants as defined in Table 9.3. All units are expressed in liters, atmospheres and degrees Kelvin. The gas constant has the value of .08206.

TABLE 9.3
Beattie-Bridgeman Constants for Common Gases

Gas	A_o	a	B_o	b	c
He	0.0216	0.05984	0.01400	0.0	0.004×10^4
Ne	0.02125	0.02196	0.02060	0.0	0.101×10^4
A	1.2907	0.02328	0.03931	0.0	5.99×10^4
N_2	1.3445	0.02617	0.05046	-.00691	4.20×10^4
O_2	1.4911	0.02562	0.04624	.004208	4.80×10^4
CO_2	5.0065	0.07132	0.10476	.07235	66.0×10^4
CH_4	2.2769	0.01855	0.05587	-.01587	12.83×10^4

The calculated coefficients of nitrogen gas at 273.15°K are listed below:

$\alpha = 22.41$

$\beta = -0.2596$

$\gamma = 0.04067$

$\delta = -1.6107 \times 10^{-5}$

When using Equation (19) to solve for an unknown volume, the following arrangement leads to an equation which is easily solved by the program entitled ROOTS.

$$-P + \frac{\alpha n}{V} + \frac{\beta n^2}{V^2} + \frac{\gamma n^3}{V^3} + \frac{\delta n^4}{V^4} = 0 \quad (24)$$

$$-PV^4 + \alpha n V^3 + \beta n^2 V^2 + \gamma n^3 V + \delta n^4 = 0 \quad (25)$$

The equation may be written as

$$A_4 V^4 + A_3 V^3 + A_2 V^2 + A_1 V + A_0 = 0 \quad (26)$$

Gas Calculations

where the coefficients are defined as follows:

$$A_4 = -P$$

$$A_3 = \alpha n$$

$$A_2 = \beta n^2$$

$$A_1 = \gamma n^3$$

$$A_0 = \delta n^4$$

1. Write a simple program to calculate the coefficients α, β, γ, and δ for a general gas at a specified temperature. Check the program by determining the coefficients of nitrogen at 273.15°K and comparing to the values listed above. Use the program to determine the coefficients for N_2 at 500°K.

2. Calculate the pressure of a mole of N_2 gas when it is confined to 200 milliliters of volume at 273.15°K. Compare to the ideal gas pressure. [Use Equation 19.]

3. Calculate the volume of a mole of nitrogen gas which is confined at 200 atmospheres and 500°K. Compare to the ideal gas volume. [Use ROOTS to determine the real root of Equation (26).]

4. Prepare a pressure-volume plot for nitrogen at 273.15°K. [Use PLOT and Equation (19).]

9.5 CRITICAL STATE

Any of a number of different equations of state may be used to study the critical state of a gas. Van der Waals equation will be chosen for this investigation. The equation for a mole of gas is written as follows:

$$P = \frac{RT}{V-b} - \frac{a}{V^2} \tag{27}$$

The constants in van der Waals equation may be defined from critical data with the following expressions:

$$a = 3P_c V_c^2 \tag{28}$$

$$b = V_c/3 \tag{29}$$

$$R = \frac{8 P_c V_c}{3 T_c} \tag{30}$$

TABLE 9.4
The Critical Constants for Some Common Gases

Gas	P_c (atm)	V_c (liters/mole)	T_c (°K)
H_2	12.8	.065	33.3
He	2.26	.0576	5.3
Ar	48.0	.0771	150.7
N_2	33.5	.090	126.1
O_2	49.7	.0744	153.4
CH_4	45.6	.0988	190.2

When the critical constants for nitrogen, as presented in Table 9.4, are substituted into Equation (28)-(30), the following van der Waals constants are obtained:

$a = 0.814$ atm liters^2 moles^{-2}

$b = .0300$ liters mole^{-1}

$R = .0638$ liter atm mole^{-1} degrees^{-1}

1. Prepare pressure-volume plots for nitrogen gas
   a. at a temperature which exceeds the critical temperature, i.e., 150°K.
   b. at the critical temperature.
   c. at a temperature below the critical temperature, i.e., 105°K.

   Make the volume the independent variable in your plot. Start the plots at a volume of approximately 0.05 liters and continue to at least 0.15 liters. [Use PLOT and Equation (27).]

2. Prepare pressure-volume plots in the critical regions for another gas found in Table 9.4.

Gas Calculations

9.6 CURVE FITTING ON EXPERIMENTAL DATA

Table 9.5 presents Boyle's Law constants (PV) for a mole of carbon dioxide gas at 373.15°K and at 471.15°K. These experimental data points may be described by a polynomial such as

$$PV = A + BP + CP^2 + DP^3 \qquad (31)$$

The coefficients in Equation (31) are easily determined by using the computer program POLEQ.

Table 9.5 also lists compressibility (Z) data for a mole of carbon dioxide at 373.15°K and 471.15°K. The experimental data points may be described in polynomial form

$$Z = A + BP + CP^2 + DP^3 \qquad (32)$$

The coefficients in Equation (32) are also determined with the computer program entitled POLEQ.

TABLE 9.5
Experimental P-V-T Data for a Mole of Carbon Dioxide

Pressure	(T = 373.15°K)		(T = 471.15°K)	
	PV	Z	PV	Z
0	30.62	1.000	38.60	1.00
50	26.87	.878		
100	22.93	.749	35.22	.912
150	19.54	.638	34.05	.882
200	18.13	.592	33.30	.863
250	18.60	.601	33.10	.859
300	19.81	.647	33.25	.861
350	21.40	.699	33.86	.877
400	23.12	.755	34.79	.901
450	24.91	.814	35.98	.932
500	26.72	.873	37.30	.966
550	28.56	.933	38.80	1.005

1. Use POLEQ to describe the data in Table 9.5 in polynomial form.
   a. PV data for CO_2 at 373.15°K.
   b. PV data for CO_2 at 471.15°K.
   c. Z data for CO_2 at 373.15°K.
   d. Z data for CO_2 and 471.15°K.

2. Use PLOT to graph each of the polynomials in the above problem.

3. Determine the volume of a mole of CO_2 gas at 75 atmospheres and $373.15°K$.

9.7 BOLTZMANN DISTRIBUTION FUNCTION

The number of molecules per element of gas velocity is given by the expression

$$\frac{dN}{dc} = 4\pi \left(\frac{m}{2\pi kT}\right)^{3/2} \exp(-\tfrac{1}{2}mc^2/kT) Nc^2 \qquad (33)$$

where k is the Boltzmann constant, T is absolute temperature, c is the velocity of the gas molecule, N is Avogadro's Number, and m is the mass of a single gas molecule. When Equation (33), is plotted over a range of molecular speeds ranging from 0 to some large velocity, the result is the classical Boltzmann distribution curve for gases. The appearance of the distribution curve changes when molecular mass or temperature is varied.

If both sides of the equation are multiplied by dc, then both sides of the equation can be integrated to produce

$$N = \int_{c_1}^{c_2} 4\pi \left(\frac{m}{2\pi kT}\right)^{3/2} \exp(-\tfrac{1}{2}mc^2/kT) Nc^2 \, dc \qquad (34)$$

The integration of the right side of equation between velocity limits c initial and c final gives the number of molecules with velocities between these limits. The integration between c = 0 and a very large c gives 6.023×10^{23} molecules. The computer integration is carried out by determining the area under the Boltzmann distribution function with the use of the program entitled AREA.

1. Plot distribution functions for each of the following:

    a. a mole of H_2 gas at $298.15°K$.

    b. a mole of H_2 gas at $1500°K$.

    c. a mole of I_2 gas at $298.15°K$.

    d. a mole of I_2 gas at $1500°K$.

    [Use PLOT and Equation (33). See Section 5.5.]

Gas Calcualtions

2. How many molecules in each of the following have speeds of 100 meters per secone or less:

   a. a mole of H_2 gas at $298.15°K$.

   b. a mole of H_2 gas at $1500°K$.

   c. a mole of I_2 gas at $298.15°K$.

   d. a mole of I_2 gas at $1500°K$.

   [Use AREA and Equation (34).]

3. A certain chemical reaction has an activation energy of 20,000 joules per mole. Therefore, individual argon atoms will have the necessary activation energy if their velocities are greater than 10^3 meters per second.

   a. Given a mole of argon gas at $298.15°K$, how many molecules possess sufficient energy for the activation?

   b. Given a mole of argon gas at $308.15°K$, how many molecules possess sufficient energy for activation?

   [Use AREA and Equation (34). Note, however, that integrations at very large values of speed produce underflow problems with the computer. See Section 5.5.]

9.8 COLLISION PROPERTIES OF GASES

The following equations describe the mean free path, the number of collisions of a single molecule per second, and the total number of collisions within a cubic meter per second, respectively:

$$L = \frac{1}{\sqrt{2}\pi\delta^2 N^*} \qquad (35)$$

$$Z_1 = \sqrt{2}\pi\delta^2 c N^* \qquad (36)$$

$$Z_{11} = \frac{1}{\sqrt{2}} \pi\delta^2 c (N^*)^2 \qquad (37)$$

If the gas of interest is nitrogen at $298°K$. the substituted values are as follows:

   δ = molecular diameter = 3.74×10^{-10} meters

   c = average speed = 470 meters per second

   N^* = number of gaseous molecules per cubic meter

The value of N* can be evaluated from the ideal gas law

$$N^* = nN = \frac{PV}{RT} N \tag{38}$$

where P, V, n, R, T are terms from the ideal gas equation and N represents Avogadro's Number. To obtain the value of N* at 298°K, consider the following numerical substitutions into Equation (39)

$$N^* = \frac{P\ (1{,}000\ \text{liters})}{.082\ (298)}\ 6.023 \times 10^{23}$$

$$N^* = 246.5 \times 10^{23}\ P$$

where P is expressed in atmospheres.

1. Write a program to compute the values of L, Z_1, and Z_{11} for nitrogen gas at 298°K and at the following pressures: 10^{-10}, 10^{-9}, 10^{-8}, ..., 10^0, 10^1, and 10^2 atmospheres.

REFERENCES

1. Van der Waals constants in Table 9.1 are from "Handbook of Chemistry and Physics", 50th edition, Chemical Rubber Publishing Company, Cleveland, Ohio, 1969, page D 135.

2. Empirical constants for the determination of virial coefficients for nitrogen gas in Table 9.2 from Samual H. Maron and David Turnbull, Journal of the American Chemical Society, 64, 44 (1942).

3. Beattie-Bridgeman, constants in Table 9.3 from James A. Beattie and Oscar C. Bridgeman, Journal of the American Chemical Society, 50, 3134 (1928).

CHAPTER 10

Molecular Energies

10.1 MOLECULAR VIBRATIONS AND ROTATIONS

In the simplest form, vibrations of diatomic molecules are described as simple harmonic oscillators, (SHO). The two atoms are pictured as masses attached by a flexible spring (bond). The quantum mechanical solution to the SHO suggests that the allowed energy levels are quantized according to the following equation:

$$\tilde{\nu} = \tilde{\nu}_c (v + \tfrac{1}{2}) \tag{1}$$

where $\tilde{\nu}$ is the allowed energy of vibration expressed in wave numbers, $\tilde{\nu}_c$ is a characteristic vibrational constant and v is a quantum number assuming integer values from 0 to infinity. The characteristic wave numbers of some common diatomic molecules are given in Table 10.1.

TABLE 10.1
Characteristic Wave Numbers for Some Diatomic Molecules

Gas	$\tilde{\nu}_c$ (cm^{-1})
H_2	4159
CO	2143
Cl_2	556
I_2	213
N_2	2331
O_2	1556
HCl	2886
HBr	2559

The simple harmonic oscillator model for vibrating molecules is adequate at low vibrational energies. The vibrations, however, display increased anharmonic motions as vibrational energies increase. An equation for the quantized anharmonic vibrational energy levels of a diatomic molecule has the form

$$\tilde{\nu} = \tilde{\nu}_c(v + \tfrac{1}{2}) - \tilde{\nu}_c x(v + \tfrac{1}{2})^2 \qquad (2)$$

where $\tilde{\nu}_c$ is a characteristic wave number, with a value slightly different from that in Equation (1), and x is the vibrational anharmonicity constant. Table 10.2 presents $\tilde{\nu}_c$ and x values for some representative gases.

TABLE 10.2
Characteristic Wave Numbers and Anharmonicity Constants For Diatomic Molecules

Gas	$\tilde{\nu}_c$ (cm^{-1})	x
H_2	4395	.02680
CO	2170	.00620
Cl_2	546.9	.00708
I_2	214.6	.000286
N_2	2360	.00613
O_2	1580	.00764
HCl	2990	.01740
HBr	2650	.01706

Molecular Energies

In the simplest form, rotating molecules are described as rigid rotators. In considering the rotational energies of molecules, the vibrations are ignored and the atoms are pictured as two masses attached to a rigid axis, the bond. The quantum mechanical solution to the rigid rotator indicates that allowed rotational energy levels are quantized according to the following equation

$$\tilde{\nu} = \frac{h}{8\pi^2 Ic} J(J + 1) \qquad (3)$$

where

$$J = 0, 1, 2, 3, 4, \cdots$$

and where $\tilde{\nu}$ is the allowed energy of an individual level expressed in wavenumbers, h is Planck's constant, c is the speed of light, and J is a quantum number. I is a moment of inertia which is defined as

$$I = \frac{m_1 m_2}{m_1 + m_2} r^2$$

where m_1 and m_2 represent the masses of the single and individual atoms within a diatomic molecule and r is the internuclear distance. The moments of inertia for some diatomic molecules are listed in Table 10.3.

TABLE 10.3
The Moments of Inertia for Selected Diatomic Molecules

Molecule	$I(\text{gr cm}^2)$	$h/8\pi^2 Ic \ (\text{cm}^{-1})$
H_2	$.4603 \times 10^{-40}$	60.81
CO	14.493×10^{-40}	1.9313
Cl_2	114.81×10^{-40}	.2438
I_2	749.42×10^{-40}	.03735
N_2	13.926×10^{-40}	2.010
O_2	19.361×10^{-40}	1.4457
HCl	2.6429×10^{-40}	10.591
HBr	3.3035×10^{-40}	8.473

As rotating molecules increase in energy, the simple rigid rotator model gives way to a model which includes centrifugal stretch. In short, the rotating molecule causes the bond to stretch somewhat. An equation for the quantized rotational energy levels for a linear molecule with centrifugal stretch has the form

$$\tilde{\nu} = B\,J(J+1) - D(J)^2(J+1)^2 \tag{4}$$

where

$$J = 0, 1, 2, 3, \cdots$$

and where $\tilde{\nu}$ is the energy in wave numbers of each different energy level, D is the centrifugal stretch constant, and B is the rotational constant defined as

$$B = \frac{h}{8\pi^2 I c}$$

The constants required in Equation (4) are summarized in Table 10.4.

TABLE 10.4
Rotational Constants for Several Diatomic Molecules

Molecule	$B(cm^{-1})$	$D(cm^{-1})$
H_2	60.81	5.200×10^{-2}
CO	1.9313	6.27×10^{-6}
Cl_2	.2438	1.88×10^{-7}
I_2	.03735	4.59×10^{-9}
N_2	2.010	5.98×10^{-6}
O_2	1.4457	4.99×10^{-6}
HCl	10.591	5.71×10^{-4}
HBr	8.473	3.72×10^{-4}

A computer program entitled ELEVL calculates the allowed energy levels according to Equations (1), (2), (3), or (4). It also calculates the differences in energy between adjacent energy levels. The program is interactive in nature. After this program has been retrieved from the computer program library, the user types RUN. The user will be asked to indicate the type of energy equation he wishes to use and he will be asked to furnish the constants which are required to complete the calculation. The program ELEVL is listed below. A sample RUN follows the listing.

```
LIST
ELEVL

10    PRINT "*********** ENERGY LEVEL PROGRAM *****************"
20    PRINT
```

Molecular Energies

```
30   PRINT "1 = SIMPLE VIBRATIONAL   - E=U*(V+.5)"
40   PRINT "2 = STRETCH VIBRATIONAL  - E=U*(V+.5)-X*U*(V+.5)↑2"
50   PRINT "3 = SIMPLE ROTATIONAL    - E=H*(J)*(J+1)/(8*P↑2*I*C)"
60   PRINT "4 = COMPLEX ROTATIONAL   - E=B*J*(J+1)-D*J↑2*(J+1)↑2"
70   PRINT "5 = EXIT FROM PROGRAM"
80   PRINT
90   PRINT "PLEASE CHOOSE 1,2,3,4, OR 5."
100  INPUT K
110  PRINT
120  GOTO K OF 140,180,220,270,660
130  GOTO 90
140  PRINT "PLEASE TYPE IN THE WAVENUMBER",
150  INPUT U
160  PRINT
170  GOTO 340
180  PRINT "INPUT ANHARMONICITY CONSTANT",
190  INPUT X
200  PRINT
210  GOTO 140
220  PRINT "THE MOMENT OF INERTIA YOU TYPE"
230  PRINT "WILL BE MULTIPLIED BY 10↑(-40)",
240  INPUT I
250  PRINT
260  GOTO 340
270  PRINT "WHAT IS THE VALUE OF 'B'",
280  INPUT B
290  PRINT
300  PRINT "PLEASE TYPE IN THE 'D' VALUE",
310  INPUT D
320  PRINT
330  GOTO 340
340  PRINT "START EVALUATION AT QUANTUM LEVEL",
350  INPUT M
360  PRINT
370  PRINT "STOP EVALUATION AT QUANTUM LEVEL",
380  INPUT L
390  PRINT
400  PRINT "........ ENERGIES WILL BE IN WAVENUMBERS OR CM↑(-1) ........"
410  PRINT
420  PRINT "************************************************************"
430  PRINT
440  PRINT "LEVEL #","E VALUE","CHANGE IN E"
450  PRINT "-------","-------","-----------"
460  PRINT
470  V=M
480  GOTO K OF 490,510,530,550
490  E=U*(V+.5)
500  GOTO 560
510  E=U*(V+.5)-X*U*(V+.5)↑2
520  GOTO 560
530  E=6.6256E+13*V*(V+1)/(3.14159↑2*8*I*2.9979E+10)
540  GOTO 560
550  E=B*V*(V+1)-D*V↑2*(V+1)↑2
560  IF V>M THEN 590
570  D5=0
580  GOTO 600
590  D5=E-E1
```

```
600   PRINT V,E,D5
610   E1=E
620   PRINT
630   V=V+1
640   IF V-1<L THEN 480
650   GOTO 80
660   END
```

```
RUN
ELEVL

*********** ENERGY LEVEL PROGRAM ******************

1 = SIMPLE VIBRATIONAL    - E=U*(V+.5)
2 = STRETCH VIBRATIONAL   - E=U*(V+.5)-X*U*(V+.5)↑2
3 = SIMPLE ROTATIONAL     - E=H*(J)*(J+1)/(8*P↑2*I*C)
4 = COMPLEX ROTATIONAL    - E=B*J*(J+1)-D*J↑2*(J+1)↑2
5 = EXIT FROM PROGRAM

PLEASE CHOOSE 1,2,3,4, OR 5.
? 1

PLEASE TYPE IN THE WAVENUMBER ? 2886

START EVALUATION AT QUANTUM LEVEL          ? 0

STOP EVALUATION AT QUANTUM LEVEL           ? 5

........ ENERGIES WILL BE IN WAVENUMBERS OR CM↑(-1) ..........

**************************************************************
```

LEVEL #	E VALUE	CHANGE IN E
0	1443	0
1	4329	2886
2	7215	2886
3	10101	2886
4	12987	2886
5	15873	2886

```
PLEASE CHOOSE 1,2,3,4, OR 5.
? 5

DONE
```

Molecular Energies 101

1. Select a molecule such as Cl_2. Calculate at least ten of the allowed vibrational energy levels by the SHO model. Calculate at least ten of the allowed vibrational levels by the anharmonic model. Compare the results of both models with special attention given to the spacings or differences between adjacent levels. [Use Equations (1) and (2), and ELEVL.]

2. Select a molecule such as I_2. Calculate at least ten of the allowed rotational energy levels by the simple rigid rotator model. Calculate at least ten of the allowed rotational levels by the centrifugal stretch model. Compare the results of both models with special attention given to the spacings or differences between adjacent levels. [Use Equations (3) and (4), and ELEVL.]

3. The HCl molecule can absorb infrared radiation to increase the vibrational energy. The selection rule requires that $\Delta v = +1$.

   a. Determine the absorption energy for an HCl molecule in the $v = 0$ state.
   b. Determine the absorption energy for an HCl molecule in the $v = 1$ state.
   c. Determine the IR absorption energies for HCl molecules in the $v = 2, 3, 4, 5, 6 \cdots$ states.

4. An HBr molecule can absorb microwave radiation to increase the rotational energy. The selection rule requires that $\Delta J = +1$.

   a. Determine the absorption energy for an HBr molecule in the $J = 0$ state.
   b. Determine the absorption energy for an HBr molecule in the $J = 1$ state.
   c. Determine the absorption energies for HBr molecules in the $J = 2, 3, 4, 5 \cdots$ states.

5. Determine the anharmonic vibrational energy levels of a molecule at large quantum levels. Observe that at high levels, the spacing between adjacent levels becomes pregressively smaller. Dissociation of the diatomic bond is said to occur when this spacing reaches zero.

   a. Determine the dissociation energy of several diatomic molecules.
   b. Show that dissociation does not occur with the SHO model.

10.2 BOLTZMANN DISTRIBUTION LAW—ROTATIONS OF MOLECULES

The Boltzmann distribution equation compares the populations of rotating molecules in two states as follows:

$$\frac{N_i}{N_j} = \frac{g_i}{g_j} \exp\left[-(\tilde{v}_j - \tilde{v}_i)hc/kT\right] \qquad (5)$$

where the subscripts i and j refer to the different rotational levels which are most frequently identified by their J values. Subscript j refers to the larger of the two levels. The g_i and g_j refer to degeneracies of the rotational levels where

$$g = 2J + 1$$

The $\tilde{\nu}_i$ and $\tilde{\nu}_j$ refer to the energies of the two rotational levels in reciprocal centimeters, h is Planck's constant with a value of 6.62×10^{-34} joule sec, c is the speed of light with a value of 3×10^{10} cm per second, k is the Boltzmann constant with a value of 1.381×10^{-23}, and T is absolute temperature. If one were to compare the number of molecules in the level $J = 1$ to the number of molecules in the level $J = 0$, Equation (5) becomes

$$\frac{N_1}{N_0} = \frac{2(1) + 1}{2(0) + 1} \exp[-(\tilde{\nu}_1 - \tilde{\nu}_0)hc/kT] \tag{6}$$

where the rotational energies described by the simple rigid rotator yield,

$$\tilde{\nu}_1 = \frac{h}{8\pi^2 Ic} 1(1 + 1) = B\, 1(1 + 1)$$

$$\tilde{\nu}_0 = \frac{h}{8\pi^2 Ic} 0(0 + 1) = B\, 0(0 + 1)$$

If one were to determine the number of molecules in the $J = 0$ level (N_0), then Equation (6) could be used to solve for the number of molecules in the $J = 1$ level (N_1). If only relative a population is desired, then the value of N_0 can be adjusted arbitrarily to a number such as a 1000, and the relative value of N_1 can be calculated. If Equation (6) is rewritten, the means of calculating the relative population of N_1 is evident;

$$N_1 = N_0 \frac{g_1}{g_0} \exp[-(\tilde{\nu}_1 - \tilde{\nu}_0)hc/kT] \tag{7}$$

Expressions similar to Equation (7) can be written for the relative populations of N_2 and N_3;

$$N_2 = N_0 \frac{g_2}{g_0} \exp[-(\tilde{\nu}_2 - \tilde{\nu}_0)hc/kT] \tag{8}$$

$$N_3 = N_0 \frac{g_3}{g_0} \exp[-(\tilde{\nu}_3 - \tilde{\nu}_0)hc/kT] \tag{9}$$

Similar equations can be written for N_4, N_5, N_6, etc.

Molecular Energies

The total rotational energy of a group of rotating molecules can be calculated by

$$E_T = hc(N_0 \tilde{v}_0 + N_1 \tilde{v}_1 + N_2 \tilde{v}_2 + \cdots) \qquad (10)$$

where E_T is expressed in joules. The average energy is simply

$$E_{ave} = \frac{E_T}{(N_0 + N_1 + N_2 + \cdots)} = \frac{E_T}{N_T} \qquad (11)$$

where N_T is the sum of all the molecules.

1. Write a computer program to determine the relative populations of rotational levels for a general gas at a given temperature. You will need to evaluate equations such as Equations (7), (8), and (9). [The moment of inertia is an extremely small number which causes underflow problems. Use the rotational constant B in your program to avoid these problems.]

2. Determine relative rotational populations of a gas such as HCl at $20°K$, at $300°K$, and at $2000°K$. [Set $N_0 = 1000$ molecules.]

3. Determine the total energy of the distribution of HCl molecules at $300°K$. [See Problem 2 and use Equation (10).]

4. Determine the average energy of the distribution of HCl molecules at $300°K$. [See Problem 3.] Compare the average rotational energy to the classical average rotational energy which has a value of kT.

5. Modify the computer program from Problem 1 to utilize rotational levels with centrifugal stretch. Determine the relative rotational populations of a gas such as HCl at $300°K$. [Set $N_0 = 1000$ molecules.]

10.3 BOLTZMANN DISTRIBUTION LAW—VIBRATIONS OF MOLECULES

The population of the v = 1 vibrational level (N_1) compared to the population of the v = 0 vibrational level (N_0) is given by the Boltzmann distribution law as

$$\frac{N_1}{N_0} = \exp[-hc\tilde{v}_c/kT] \qquad (12)$$

The differene in energy between the v = 0 and v = 1 level is expressed in terms of the simple harmonic oscillator model. When the value of N_0 is known, this equation may be rewritten as

$$N_1 = N_0 \exp[-x] \qquad (13)$$

where

$$x = \frac{hc\tilde{\nu}_c}{kT} \qquad (14)$$

Notice that the degeneracies for vibrational levels are equal to one and can, therefore, be neglected. The population of the v = 2 vibrational level (N_2) is given by

$$N_2 = N_0 \exp(-2x) \qquad (15)$$

Similar expressions can be written for populations in the other quantum states. It can be shown that the population in the N_0 level is determined by

$$N_0 = N_T [1 - \exp(-x)] \qquad (16)$$

N_T represents the total number of molecules in a grouping and usually is taken as 6.023×10^{23} molecules.

Part of the total vibrational energy is zero-point energy, defined as $\tfrac{1}{2}hc\tilde{\nu}_c$. The remainder of the energy is classified as thermal vibrational energy. The vibrational thermal energy of a distribution of diatomic molecules is

$$E_T = N_0(0) + N_1(hc\tilde{\nu}_c) + N_2(2hc\tilde{\nu}_c) + N_3(3hc\tilde{\nu}_c) + \cdots \qquad (17)$$

The average thermal vibrational energy of a molecule within the distribution of vibrating molecules is

$$E_{ave} = \frac{E_T}{(N_0 + N_1 + N_2 + \cdots)} = \frac{E_T}{N_T} \qquad (18)$$

1. Write a computer program to determine the relative populations of vibrational levels for a general diatomic gas at a given temperature. You will need to define x by Equation (14), define N_0 by Equation (16) and to calculate N_1, N_2, $N_3 \cdots$ by Equations such as (13) and (15).

2. Determine relative vibrational populations of a gas such as HCl at $20°K$, $300°K$, and $2000°K$.

3. Determine relative vibrational populations for each of the following gases at $300°K$: I_2, O_2, H_2. Compare the populations.

4. Determine the thermal vibrational energy of a mole of O_2 molecules at $300°K$.

5. Determine the average thermal vibrational energy of each O_2 molecule at $300°K$. Compare to the classical average vibrational energy of kT.

10.4 FIRST EINSTEIN FUNCTION

The total thermal vibrational energy of a mole of diatomic molecules can be determined by evaluating Equation (17) or it can be determined by using

$$E_T = RT \frac{x}{e^x - 1} \qquad (19)$$

where E_T is the molar thermal vibrational energy, R is a gas constant frequently expressed in calories or joules, T is the absolute temperature, and x is defined as

$$x = \frac{hc\tilde{\nu}_c}{kT} \qquad (20)$$

where h is Planck's constant, c is the speed of light, $\tilde{\nu}_c$ is the characteristic wavenumber of the vibrations and k is the Boltzmann constant. When $\tilde{\nu}_c$ is expressed in reciprocal centimeters, then the speed of light is expressed in c.g.s. units. The part of Equation (19) involving x is named the first Einstein function, $f_1(x)$; therefore,

$$E_T = RT\, f_1(x) \qquad (21)$$

where

$$f_1(x) = \frac{x}{e^x - 1} \qquad (22)$$

1. Write a computer program to evaluate the first Einstein function as a function of x. Use the program to prepare a table of Einstein values for x ranging from x = 0 to x = 20. The table may be used much like a table of cosine function.

2. PLOT the first Einstein function over a range of x from 0 to 10. [Use PLOT and Equation (22). See Section 5.5.]

3. Determine the thermal vibrational energy of a mole of O_2 gas at 100°K, 300°K, and 2000°K. [Write your own program using Equations (19) and (20). See Table 10.1 for the characteristic wave numbers.]

4. Determine the thermal vibrational energy for a mole of each of the following gases at 500°K: H_2, HCl, CO, Cl_2, and I_2. [See Table 10.1 for the characteristic wave numbers.]

5. A mole of N_2 molecules has a thermal vibrational energy of 4000 joules per mole. What is the temperature of the N_2 gas?

10.5 MOLECULAR ENERGIES OF POLYATOMIC MOLECULES

The total thermal energy of a mole of gaseous molecules is the sum of the translational, the rotational and the vibrational energies of the molecule. Molar translational energies are given by

$$E_{trans} = \frac{3}{2} RT \tag{23}$$

Molar rotational energies at room temperature and above are generally classical in nature and expressed as

$$E_{rot} = \frac{3}{2} RT \quad \text{(non-linear)} \tag{24}$$

$$E_{rot} = \frac{2}{2} RT \quad \text{(linear)} \tag{25}$$

Finally, molar thermal vibrational energies are expressed by

$$E_{vib} = \sum_{i=1}^{3n-6} RT \, f_1(x_i) \quad \text{(non-linear)} \tag{26}$$

$$E_{vib} = \sum_{i=1}^{3n-5} RT \, f_1(x_i) \quad \text{(linear)} \tag{27}$$

where the vibrational contribution is the sum of contributions from all fundamental vibrations. The total thermal molecular energy is therefore

$$E_T = E_{trans} + E_{rot} + E_{vib} \tag{28}$$

TABLE 10.5
CHARACTERISTIC WAVE NUMBERS FOR POLYATOMIC MOLECULES

Molecule	$\tilde{\nu}_c$ (cm^{-1})
H_2O	3756, 3652, 1595
CO_2	2349, 1340, 667, 667
SO_2	1361, 1151, 524
C_6H_6	992, 3062, 606, 606, 1178, 1178, 1596, 1596, 845, 845, 3047, 3047, 1010, 3060, 406, 406, 1037, 1037, 1485, 1485, 850, 850, 3080, 3080, 1648, 1326, 664, 671, 1648, 1326

Molecular Energies 107

1. Determine the total molar thermal energy of a simple polyatomic molecule such as CO_2, at $500°K$.

2. Determine the total molar thermal energy of a molecule like benzene at $500°K$.

10.6 HEAT CAPACITIES OF IDEAL GASES

At temperatures of $300°K$ or greater, the heat capacity contributions due to translations and rotations are classical in nature. The translational heat capacity including pressure-volume work is

$$C_{p,trans} = \frac{5}{2} R \tag{29}$$

The rotational heat capacity for a linear molecule is

$$C_{p,rot} = \frac{2}{2} R \tag{30}$$

Vibrational heat capacities do not become classical until high temperatures are attained. The heat capacity contribution from one vibrational mode is given by

$$C_{p,vib} = R \frac{x^2 e^x}{(e^x - 1)^2} \tag{31}$$

where x is defined by Equation (20). The part of Equation (31) involving x is called the second Einstein function, $f_2(x)$; therefore,

$$C_{p,vib} = R\, f_2(x) \tag{32}$$

where

$$f_2(x) = \frac{x^2 e^x}{(e^x - 1)^2} \tag{33}$$

1. Calculate the total heat capacity for a mole of HCl gas over the range of $300°$ to $2000°K$. Compare the heat capacity at $2000°K$ with the classical heat capacity.

2. Calculate the total heat capacity for a mole of Cl_2 gas over the range of $300°$ to $2000°K$. Compare the heat capacity at $2000°K$ with the classical heat capacity.

3. PLOT the value of the second Einstein function over the range of characteristic wave numbers from 200 to 4000 reciprocal centimeters when the temperature is $2000°K$.

4. PLOT the value of the second Einstein function over the temperature range of $300°$ to $2000°K$ when the characteristic wave number is a constant 1556 reciprocal centimeters.

REFERENCES

1. The vibrational and rotational constants for the diatomic molecules in Tables 10.1, 10.2, 10.3, and 10.4 are based on constants found in "Molecular Spectra and Molecular Structures", Volume I, Spectra of Diatomic Molecules, by G. Hertzberg; D. Van Nostad Company, Inc.; Princeton, New Jersey (1965); Table 39.

2. The vibration constants for the triatomic molecules in Table 10.5 are from "Molecular Spectra and Molecular Structure", Volume III, Electronic Spectra and Electronic Structure of Polyatomic Molecules, by G. Hertzberg; D. Van Nostad Company, Inc.; Princeton, New Jersey (1965); Table 62.

3. The vibrational constants for benzene are from Kovner, M. A., Journal of Experimental and Theoretical Physics, 26, 598 1954; Lord, R. C., Jr. and Andrews, D. H., Journal of Physical Chemistry, 41, 149 (1937).

CHAPTER 11

Classical Thermodynamics

11.1 EXPERIMENTAL HEAT CAPACITIES

Molar heat capacities find constant use in the study of classical thermodynamics. The heat capacities may be determined experimentally and then expressed in polynomial form such as

$$C_p^o = a + bT + cT^2 + dT^3 + \cdots \qquad (1)$$

where C_p^o is the molar heat capacity for a given substance at a constant one atmosphere of pressure, T is absolute temperature and a, b, c, d, $\cdots$ are experimentally determined coefficients. In general, the sets of coefficients are different for the solid, the liquid and the gas phase. Table 11.1 lists coefficients for a number of common substances. These coefficients will be useful in solving problems throughout the entire chapter.

TABLE 11.1
Molar Heat Capacities at one Atmosphere Pressure
In Joules Per Mole Per Degree Over 300-1500°K Range

$$C_p^o = a + bT + cT^2$$

Compound	a	b x 10^3	c x 10^7
$H_2(g)$	29.068	-0.8365	20.12
$O_2(g)$	25.505	13.613	-42.56
$N_2(g)$	26.986	5.910	-3.376
$CO(g)$	26.539	7.684	-11.72
$CO_2(g)$	26.762	42.651	-147.85
$H_2O(g)$	30.074	9.931	8.720
$H_2O(l)$	75.48	0.00	0.00
$C(s)$	6.106	19.74	-50.81

1. Listed below are some experimental molar heat capacities for CH_4 at a constant one atmosphere pressure. Determine the coefficients for Equation (1) which best fit the experimental data. [Use POLEQ.]

Temperature (°K)	C_p joules/degree
300	35.15
400	41.39
500	47.29
600	52.84
700	58.04
800	62.89
900	67.40
1000	71.55
1100	75.36
1200	78.82
1300	81.93
1400	84.69
1500	87.10

2. Graph the molar heat capacity for O_2 over the temperature range 300 to 1500°K. [Use PLOT.]

3. At what temperature does N_2 gas have a molar heat capacity of 34.55 joules per mole per degree? [Use ROOTS.]

Classical Thermodynamics

11.2 HEAT CAPACITIES—MOLECULAR MODEL

The total molar heat capacity of a gas is considered to contain contributions from translational motions, from rotation motions, from vibrational motions, and from pressure-volume work as follows:

$$Cp = \frac{3}{2} R + \left(\frac{2}{2} \text{ or } \frac{3}{2}\right) R + \sum R \frac{x^2 e^x}{(e^x-1)^2} + R \qquad (2)$$

where R is a gas constant expressed in calories or joules, and x is defined as

$$x = \frac{hc\tilde{\nu}_c}{kT} \qquad (3)$$

The first term in Equation (2) reflects translation heat capacity, the second term reflects linear (2/2) or non-linear (3/2) rotational heat capacities, the third term reflects vibrational heat capacities summed over all the fundamental modes, and the final R reflects the pressure-volume work according to the ideal gas law. The portion of Equation (2) involving x is designated as the second Einstein function, $f_2(x)$; therefore,

$$f_2(x) = \frac{x^2 e^x}{(e^x-1)^2} \qquad (4)$$

1. Write a computer program to evaluate the second Einstein function at different values of x. Use the program to prepare a table of Einstein values for x ranging from x = 1 to x = 15.

2. PLOT the second Einstein function over a range of x from 0.001 to 15. [Use PLOT and Equation (4).]

3. Determine the vibrational heat capacity of a mole of each of the following gases at 500°K: H_2, HCl, CO, Cl_2, and I_2. [Write your own program using the vibrational portion of Equation (2) and Equation (3). See Table 10.1 for the characteristic wave numbers.]

4. Determine the total heat capacity for a mole of CO_2 gas at 300, 400, 500, ··· 1500°K. [See Table 10.5 for the fundamental wave numbers. Use Equation (2).]

5. Use the temperatures and heat capacities of CO_2 gas from problem 4 to determine the coefficients in the equation

$$C_p = a + bT + cT^2$$

[Use POLEQ.]

11.3 GAS EXPANSIONS AND COMPRESSIONS

An ideal and monatomic gas which is confined to a piston and cylinder arrangement may be expanded by four different methods. The same gas may also be compressed by four different methods. The basic equations necessary for the calculation of q, w, and ΔE for each method are summarized below.

Irreversible and Isothermal Expansion or Compression

$$w = P(V_2 - V_1) \tag{5}$$

Reversible and Isothermal Expansion or Compression

$$w = nRT \ln \frac{V_2}{V_1} \tag{6}$$

Irreversible and Adiabatic Expansion or Compression

$$nC_v(T_2 - T_1) = -P(V_2 - V_1) \tag{7}$$

Reversible and Adiabatic Expansion or Compression

$$C_v \ln \frac{T_2}{T_1} = -R \ln \frac{V_2}{V_1} \tag{8}$$

The C_v for an ideal and monatomic gas is 3/2 R. When the final pressure is indicated rather than the final volume, the ideal gas substitution is necessary where

$$V_2 = \frac{nRT_2}{P_2}$$

1. A piston and cylinder arrangement contains one liter of ideal and monatomic gas at 10 atmospheres pressure and 298.15°K. Calculate the w, q, and ΔE for the expansion of the gas to a final pressure of one atmosphere by each of the methods listed below. Also calculate the final temperatures and the final volumes. Compare the results for the different methods.

    a. Irreversible and isothermal
    b. Reversible and isothermal
    c. Irreversible and adiabatic
    d. Reversible and adiabatic

2. A piston and cylinder arrangement contains 10 liters of ideal and monatomic gas at one atmosphere pressure and 298.15°K. Calculate the w, q, and ΔE for the compression of the gas to a final pressure of 10 atmospheres by each of

Classical Thermodynamics

methods listed below. Calculate also the final temperatures and final volumes. Compare all results.

    a.   Irreversible and isothermal

    b.   Reversible and isothermal

    c.   Irreversible and adiabatic

    d.   Reversible and adiabatic

11.4 HEATING GASES

The amount of heat required to increase the temperature of a mole of substance is given by

$$\Delta H^o = \int_{T_1}^{T_2} (a + bT + cT^2) \, dT \tag{9}$$

where ΔH^o is the heat required in joules, the a, b, and c are the coefficients in the heat capacity polynomial, and T is absolute temperature. The integral is evaluated from an initial temperature T_1 to a final temperature T_2. The value of the integral is easily obtained with the use of a program entitled AREA.

1. Determine the heat required to increase the temperature of a mole of $CO_2(g)$ from $298.15°K$ to $900°K$.

2. Determine the heat required to increase the temperature of a mole of $H_2(g)$ from $298.15°K$ to $900°K$.

3. Determine the heat lost when a mole of $CO(g)$ is cooled from a $900°K$ to $298.15°K$.

4. Determine the heat lost when a mole of $H_2O(g)$ is cooled from a $900°K$ to $298.15°K$.

11.5 ENTHALPIES OF REACTION

Consider the following chemical reaction as a useful example for several thermodynamic calculations:

$$CO(g) + H_2O(g) \rightarrow CO_2(g) + H_2(g)$$

The standard heat of this reaction (ΔH^o) at 298.15°K is obtained by substituting standard enthalpies of formation at 298.15°K and one atmosphere pressure, as found in Table 11.2, into the equation

$$\Delta H^o = (\Delta H_f^o)_{CO_2} + (\Delta H_f^o)_{H_2} - (\Delta H_f^o)_{H_2O} - (\Delta H_f^o)_{CO} \quad (10)$$

$$= -393,510 + 0 - (-241,830) - 110,520)$$

$$= -41,160 \text{ joules}$$

TABLE 11.2
Enthalpies of Formation in Joules Per Mole
at 298.15°K and one atmosphere pressure

Gas	ΔH_f^o (joules)
C(s)	0.0
H_2(g)	0.0
O_2(g)	0.0
N_2(g)	0.0
CO(g)	-110,520
CO_2(g)	-393,510
H_2O(l)	-285,840
H_2O(g)	-241,830

The heat of this same reaction at an elevated temperature can be determined with the aid of Figure 11.1.

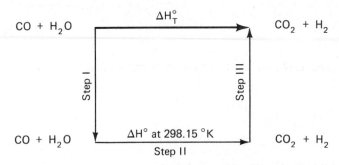

Figure 11.1 The heat of reaction at elevated temperature

The desired heat of the reaction at the elevated temperature, ΔH_T^o, may be found by summing Steps I, II, and III. Step I involves cooling the CO and H_2O to 298.15°K,

Step II is the standard ΔH° for the reaction as determined above, and Step III involves the heating of the products CO_2 and H_2 to the elevated temperature.

1. Calculate the ΔH_T° for the reaction at $900°K$ using the method described in Figure 11.1. [Use calculated results from the problems in Section 11.4.]

2. Calculate the ΔH_T° for the reaction at $1500°K$ using the method described in Figure 11.1.

A second method for calculating the heat of the same reaction at an elevated temperature uses a mathematical form of Figure 11.1.

$$\Delta H_T^\circ = \Delta H^\circ + \int_{298.15}^{T} (\Delta a + \Delta bT + \Delta cT^2)\, dT \qquad (11)$$

where Δa, Δb, and Δc are determined from heat capacity coefficients as follows:

$$\Delta a = a_{CO_2} + a_{H_2} - a_{CO} - a_{H_2O}$$

$$\Delta b = b_{CO_2} + b_{H_2} - b_{CO} - b_{H_2O}$$

$$\Delta c = c_{CO_2} + c_{H_2} - c_{CO} - c_{H_2O}$$

The substitution of coefficients from Table 11.1 yields

$$\Delta a = -0.783$$
$$\Delta b = 24.200 \times 10^{-3}$$
$$\Delta c = -124.73 \times 10^{-7}$$

3. Calculate the ΔH_T° for the reaction at $900°K$ using Equation (11).
4. Calculate the ΔH_T° for the reaction at $1500°K$ using Equation (11).
5. Use data from Tables 11.1 and 11.2 to calculate the ΔH_T° at $373.15°K$ for the vaporization of water.

$$H_2O(\ell) \rightarrow H_2O(g)$$

A third method for calculating the heat of the reaction at an elevated temperature involves the integrated form of Equation (11).

$$\Delta H_T^\circ = \Delta aT + \frac{\Delta bT^2}{2} + \frac{\Delta cT^3}{3} + I \qquad (12)$$

where I for the carbon monoxide and water vapor reaction is $-41,892$ joules per mole.

6. PLOT the ΔH_T° for the reaction from 300 to $1500°K$.

7. Calculate the ΔH_T^o for the reaction at 900°K and 1500°K using Equation (12).

The fourth method for calculating the heat of reaction at an elevated temperature consists of calculating the heats of formation for each of the reactants and products at the elevated temperature by the equation

$$(\Delta H_f^o)_T = \Delta H_f^o + \int_{298.15}^T \Delta C_p \, dT \qquad (13)$$

To calculate the $(\Delta H_f^o)_T$ of CO_2, for example, the ΔH_f^o for CO_2 from Table 11.2 is substituted. The ΔC_p for the formation reaction of CO_2 is calculated, and Equation (13) is evaluated with the aid of program AREA. Similar substitutions are required to calculate the heat of formation of $CO(g)$ and of $H_2O(g)$ at elevated temperatures. The heat of formation at the elevated temperature for the element $H_2(g)$ is zero. The heat of the reaction is then given by

$$\Delta H_T^o = (\Delta H_f^o)_{T,CO_2} + (\Delta H_f^o)_{T,H_2} - (\Delta H_f^o)_{T,CO} - (\Delta H_f^o)_{T,H_2O} \qquad (14)$$

8. Calculate the ΔH_T^o for the reaction at 900°K using Equation (14).
9. Calculate the ΔH_T^o for the reaction at 1500°K using Equation (14).

11.6 THIRD LAW ENTROPY

Consider the determination of a third law entropy of a substance such as nitrogen, which is a gas at room temperature. The entropy at 0°K is defined as zero by the third law of thermodynamics. Knowledge of the heat capacities, the transition temperatures and the transition enthalpies allows an approximate determination of the entropy at 298.15°K. Data for nitrogen is summarized below:

a. The solid phase I of N_2 has a heat capacity as follows:

   $C_p = 0.00 + 0.763T$

b. The crystalline form of N_2 changes to solid phase II at 35.61°K with a transition heat of 228.9 joules per mole.

c. Solid phase II of N_2 has a heat capacity

   $C_p = 21.78 + 0.3968T$

d. Solid nitrogen melts at 63.14°K with a heat of melting of 720.9 joules per mole.

e. The liquid heat capacity is approximately

   $C_p = 48.66 + 0.1102T$

Classical Thermodynamics

f. Liquid nitrogen boils at 77.32°K with a heat of vaporization of 5535 joules per mole.

g. The approximate heat capacity of gaseous nitrogen at temperatures below 300°K is approximately 29.10 joules per mole per degree.

An approximate entropy of nitrogen at 298.15°K is determined by summing the entropy changes for each of the steps outlined above. Whenever a given phase is increased in temperature the entropy is determined by

$$\Delta S^{o} = \int_{T_1}^{T_2} \frac{C_p}{T} dT \tag{15}$$

The value of the integral is determined with the program AREA. Whenever a phase change occurs the entropy is given by

$$\Delta S^{o} = \frac{\Delta H_{trans}}{\Delta T_{trans}} \tag{16}$$

1. Calculate an approximate third law entropy of N_2 gas at 298.15°K. Your result will not include the usual statistical calculations at the near absolute zero temperature; nor will not include the usual correction for gas imperfections.

2. Given S^o for N_2 at 298.15°K as 191.49 joules per mole per degree. Calculate the third law entropy of N_2 at 1500°K. [Use the heat capacity from Table 11.1.]

The third law entropies at 298.15°K and at one atmosphere pressure, as found in Table 11.3, will be useful in calculations to determine the third law entropies of elements or compounds at elevated temperatures.

TABLE 11.3

Entropies in Joules Per Mole Per Degree at 298.15°K and one atmosphere pressure

Gas	S^o (joules degrees^{-1})
$H_2(g)$	130.59
$O_2(g)$	205.03
$N_2(g)$	191.49
$CO(g)$	197.91
$CO_2(g)$	213.64
$H_2O(l)$	69.94
$H_2O(g)$	188.72
$C(s)$	5.69

3. Determine the third law entropy for CO(g) at 900°K and 1500°K.
4. Determine the third law entropy for H_2O(g) at 900°K and 1500°K.
5. Determine the third law entropy for CO_2(g) at 900°K and 1500°K.
6. Determine the third law entropy for H_2(g) at 900°K and 1500°K.

11.7 ENTROPIES FOR REACTIONS

Consider, again, the chemical reaction

$$CO(g) + H_2O(g) \rightarrow CO_2(g) + H_2(g)$$

The $\Delta S°$ for the reaction at standard conditions is given by Equation (17) and the entropy values from Table 11.3.

$$\Delta S° = S°_{CO_2} + S°_{H_2} - S°_{CO} - S°_{H_2O} \tag{17}$$

$$= 213.64 + 130.59 - 197.91 - 188.72$$

$$= -42.40 \text{ joules per degree}$$

The $\Delta S°$ for the same reaction at elevated temperatures may be determined by using

$$\Delta S°_T = \Delta S° + \int_{298.15}^{T} \left[\frac{\Delta a + \Delta bT + \Delta cT^2}{T}\right] dT \tag{18}$$

where $\Delta S°$ is the entropy for the reaction at 298.15°K and the Δa, Δb, and Δc have values as determined in Section 11.5.

1. Calculate $\Delta S°_T$ for the reaction at 900°K using Equation (18).
2. Calculate $\Delta S°_T$ for the reaction at 1500°K using Equation (18).

The same entropies for the reaction at an elevated temperature may be determined by using the integrated form of Equation (18).

$$\Delta S°_T = \Delta a \ln T + \Delta bT + \frac{\Delta c}{2} T^2 + I \tag{19}$$

where I has the value of -44.60 for the carbon monoxide and water vapor reaction.

3. PLOT the $\Delta S°_T$ for the reaction from 300 to 1500°K.
4. Calculate the $\Delta S°_T$ values for the reaction at 900 and 1500°K.

Classical Thermodynamics 119

Another method for calculating the ΔS_T^o for the reaction is by using third law entropies at elevated temperatures where

$$\Delta S_T^o = (S_T^o)_{CO_2} + (S_T^o)_{H_2} - (S_T^o)_{CO} - (S_T^o)_{H_2O} \qquad (20)$$

5. Calculate ΔS_T^o for the reaction at $900^\circ K$ using Equation (20).
6. Calculate ΔS_T^o for the reaction at $1500^\circ K$ using Equation (20). [See Questions 3, 4, 5, and 6 from Section 11.6.]

11.8 FREE ENERGIES OF REACTIONS

Several different methods may be employed to determine the free energy of the reaction

$$CO(g) + H_2O(g) \rightarrow CO_2(g) + H_2(g)$$

The expression for ΔG^o

$$\Delta G^o = \Delta H^o - T\Delta S^o \qquad (21)$$

may be used to calculate the free energy of the reaction.

1. Calculate ΔG^o for the reaction at $298.15^\circ K$. [Use calculated results from Sections 11.5 and 11.7.]
2. Calculate ΔG_T^o for the reaction at $900^\circ K$.
3. Calculate ΔG_T^o for the reaction at $1500^\circ K$.

TABLE 11.4
Free Energies of Formation at $298.15^\circ K$ and one atmosphere pressure

Substance	ΔG_f^o (joules per mole)
$C(s)$	0
$H_2(g)$	0
$O_2(g)$	0
$N_2(g)$	0
$CO(g)$	-137,270
$CO_2(g)$	-394,380
$H_2O(l)$	-237,190
$H_2O(g)$	-228,590

The ΔG^o at 298.15°K for the reaction may also be obtained by employing Equation (22) and the free energies from Table 11.4.

$$\Delta G^o = (\Delta G_f^o)_{CO_2} + (\Delta G_f^o)_{H_2} - (\Delta G_f^o)_{CO} - (\Delta G_f^o)_{H_2O} \qquad (22)$$

$$= -394,380 - 0 - (-137,270) - (-228,590)$$

$$= -28,520$$

For a second method of calculating ΔG_T^o for the reaction at elevated temperatures consider the differential equation

$$\left(\frac{\delta G}{\delta T}\right)_P = -\Delta S \qquad (23)$$

When this differential and Equation (19) are applied to a chemical reaction, the following equation is obtained:

$$\Delta G_T^o = \Delta C^o - \int_{298.15}^{T} (\Delta a \ln T + \Delta b T + \frac{\Delta c}{2} T^2 + I) \, dT \qquad (24)$$

where ΔG^o is defined above and Δa, Δb, and Δc are defined in Section 11.5. The value of I in Equation (24) is -44.60 joules per mole per degree for the carbon monoxide and water vapor reaction.

4. Calculate ΔG_T^o for the reaction at 900°K using Equation (24).
5. Calculate ΔG_T^o for the reaction at 1500°K using Equation (24).

An integrated Gibbs-Helmholtz equation is another possible equation for the calculation of free energy at elevated temperatures.

$$\Delta G_T^o = -\Delta a T \ln T - \frac{\Delta b}{2} T^2 - \frac{\Delta c}{6} T^3 + I + I'T \qquad (25)$$

I has a value of -41,891 joules per mole and I' has a value of 43.816 joules per mole per degree for the considered reaction.

6. PLOT ΔG_T^o for temperatures ranging from 300 to 1500°K.
7. Determine ΔG_T^o at 900°K and 1500°K.

11.9 FREE ENERGY AND PRESSURE

Consider mixing $CO(g)$, $H_2O(g)$, $CO_2(g)$ and $H_2(g)$ so that the initial partial pressures are P_{CO}, P_{H_2O}, P_{CO_2}, and P_{H_2}. Assume that the mixture of gases will then react according to the equation

$$CO(g) + H_2O(g) \rightarrow CO_2(g) + H_2(g)$$

The free energies of each of the substances will differ from those found in thermodynamic tables because the pressures differ from one standard atmosphere. With the aid of the ideal gas law, the free energies of each of the gases in the initial mixture may be calculated as follows:

$$\text{Free energy for CO} = \Delta G_f^o + RT \ln P_{CO} \qquad (26)$$

$$\text{Free energy for } H_2O = \Delta G_f^o + RT \ln P_{H_2O} \qquad (27)$$

$$\text{Free energy for } CO_2 = \Delta G_f^o + RT \ln P_{CO_2} \qquad (28)$$

$$\text{Free energy for } H_2 = \Delta G_f^o + RT \ln P_{H_2} \qquad (29)$$

The free energy for the entire reaction is easily condensed to

$$\Delta G = \Delta G^o + RT \ln \frac{P_{CO_2} P_{H_2}}{P_{CO} P_{H_2O}} \qquad (30)$$

where ΔG^o (or ΔG_T^o) is defined in Section 11.8. Assume that initially the ΔG is a negative number. Carbon monoxide and water vapor will therefore be converted into carbon dioxide and hydrogen gas. As the reaction continues, the partial pressures of the CO_2 and H_2 increase, while the pressures of the CO and H_2O decrease. While the partial pressures are changing, the calculated ΔG moves closer to a zero value. When ΔG reaches a zero value, the reaction is at equilibrium and the reaction stops.

A short computer program entitled GFORRX will calculate the ΔG of the reaction as it progresses toward equilibrium. The program user will be asked to input the absolute temperature and the initial partial pressures of the CO, H_2O, CO_2 and H_2. The operator will also be asked to control the progress of the reaction by inputting a small x which represents the decrease in the partial pressure of CO. The x value also represents the decrease in the H_2O partial pressure, it represents the increase in the CO_2 partial pressure, and the increase in the H_2 partial pressure. The value of ΔG will be calculated by the computer program. On each new iteration the operator will slowly increase the value of x until the value of ΔG reaches zero, the point of equilibrium. The ratio of partial pressure when ΔG reaches zero is defined as the equilibrium constant.

$$K_p = \frac{P_{CO_2} P_{H_2}}{P_{CO} P_{H_2O}} \qquad (31)$$

```
LIST
GFORRX

5     DEF FNF(T1)=-A1*T1*LOG(T1)-B1/2*T1↑2-C1/6*T1↑3+D1+D2*T1
10    LET A1=-.783
15    LET B1=.0242
20    LET C1=-1.2473E-05
25    LET D1=-41891.
30    LET D2=43.816
35    DEF FNG(T1)=FNF(T1)+8.314*T1*LOG((P3+X1)*(P4+X1)/((P1-X1)*(P2-X1)))
40    PRINT "YOU ARE RUNNING THE REACTION:   CO + H2O = CO2 + H2"
45    PRINT
50    PRINT "AT WHAT TEMPERATURE IS IT TAKING PLACE";
55    INPUT T1
60    PRINT
65    PRINT "STANDARD FREE ENERGY FOR THIS REACTION IS",FNF(T1)
70    PRINT
75    PRINT "WHAT ARE THE INITIAL PARTIAL PRESSURES FOR THE FOLLOWING:"
80    PRINT
85    PRINT "CO",
90    INPUT P1
95    PRINT "H2O",
100   INPUT P2
105   PRINT "CO2",
110   INPUT P3
115   PRINT "H2",
120   INPUT P4
125   PRINT
130   X1=0
135   PRINT "FREE ENERGY FOR THE REACTION IS NOW",FNG(T1)
140   PRINT
145   PRINT "INPUT CHANGE IN PRESSURE (DELTA P) OR ZERO (0.) TO STOP"
150   PRINT
155   PRINT
160   PRINT "DELTA P";
165   INPUT X1
170   PRINT
175   IF X1>0 THEN 210
180   IF X1<0 THEN 190
185   GOTO 265
190   IF ABS(X1)>P3 MIN P4 THEN 200
195   GOTO 225
200   PRINT "THIS REVERSAL WOULD LEAD TO NEGATIVE PRODUCT PRESSURES"
205   GOTO 150
210   IF X1<P1 MIN P2 THEN 225
215   PRINT "THIS CHANGE WOULD USE UP MORE THAN THE INITIAL PRODUCTS"
220   GOTO 160
225   PRINT
230   PRINT "NEW FREE ENERGY FOR REACTION =",FNG(T1)
235   PRINT
240   PRINT "P OF CO",P1-X1
245   PRINT "P OF H2O",P2-X1
250   PRINT "P OF CO2",P3+X1
255   PRINT "P OF H2",P4+X1
260   GOTO 150
265   END
```

RUN
GFORRX

YOU ARE RUNNING THE REACTION: $CO + H_2O = CO_2 + H_2$

AT WHAT TEMPERATURE IS IT TAKING PLACE? 900

STANDARD FREE ENERGY FOR THIS REACTION IS   -5948.48

WHAT ARE THE INITIAL PARTIAL PRESSURES FOR THE FOLLOWING:

```
CO              ? 1
H2O             ? 1
CO2             ? 1
H2              ? 1
```

FREE ENERGY FOR THE REACTION IS NOW   -5948.49

INPUT CHANGE IN PRESSURE (DELTA P) OR ZERO (0.) TO STOP

DELTA P? .25

NEW FREE ENERGY FOR REACTION = 1696.12

```
P OF CO           .75
P OF H2O          .75
P OF CO2         1.25
P OF H2          1.25
```

DELTA P? .20

NEW FREE ENERGY FOR REACTION = 119.379

```
P OF CO           .8
P OF H2O          .8
P OF CO2         1.2
P OF H2          1.2
```

DELTA P? .196

NEW FREE ENERGY FOR REACTION =-5.22559

```
P OF CO           .804
P OF H2O          .804
P OF CO2         1.196
P OF H2          1.196
```

DELTA P?0

DONE

1. A container and its contents are maintained at 900°K. Gases initially introduced into the container have partial pressures as follows: CO, 1.0 atmosphere; H_2O, 1.0 atmosphere; CO_2, 1.0 atmosphere; H_2, 1.0 atmosphere. Use the program GFORRX to calculate the equilibrium partial pressures. Also calculate the equilibrium constant as defined in Equation (31).

2. A container and contents are maintained at 1500°K. The initial partial pressures of CO, H_2O, CO_2, and H_2 are one atmosphere of each. Calculate the equilibrium partial pressures and the equilibrium constant. [Use GFORRX.]

3. A container and contents are maintained at 900°K. The initial partial pressures of CO, H_2O, CO_2, and H_2 are 2.0, 1.0, 0.2, and 0.1 atmospheres respectively. Calculate the equilibrium partial pressures and the equilibrium constant. [Use GFORRX.]

4. A container and contents are maintained at 1500°K. The initial partial pressures of CO, H_2O, CO_2, and H_2 are 2.0, 1.0, 0.2, and 0.1 atmospheres respectively. Calculate the equilibrium partial pressures and the equilibrium constant. [Use GFORRX.]

11.10 EQUILIBRIUM CONSTANTS

The free energy of a reaction reaches zero at equilibrium, therefore, Equation (30) may be rewritten at equilibrium as

$$\Delta G^\circ = -RT \ln K_p \qquad (32)$$

The value of ΔG° (or ΔG_T°) is sufficient to determine the value of the equilibrium constant.

1. Calculate the equilibrium constant of the following reaction at 300, 400, 500, 600, ... 1500°K. [Use Equation (32) and information from the previous sections.]

    $$CO(g) + H_2O \rightarrow CO_2(g) + H_2(g)$$

2. Calculate the equilibrium constant of the following reaction at 400, 500, 600, ... 1500°K. [Use any information from the previous sections.]

    $$H_2O(g) \rightarrow H_2(g) + \tfrac{1}{2} O_2(g)$$

Classical Thermodynamics

11.11 EQUILIBRIUM CONCENTRATIONS

If the equilibrium constant for a given chemical reaction is known, the chemist can predict the equilibrium concentrations. Consider again the reaction

$$CO(g) + H_2O(g) \rightarrow CO_2(g) + H_2(g)$$

A mixture with initial partial pressures P_{CO}, P_{H_2O}, P_{CO_2}, and P_{H_2} follows the equilibrium expression

$$K_p = \frac{(P_{CO_2} + X)(P_{H_2} + X)}{(P_{CO} - X)(P_{H_2O} - X)} \tag{33}$$

where X represents the loss in partial pressure for CO. The equation is easily rearranged so that

$$(K_p - 1)X^2 + (-K_p P_{CO} - K_p P_{H_2O} - P_{CO_2} - P_{H_2})X$$
$$+ (K_p P_{CO} P_{H_2O} - P_{CO_2} P_{H_2}) = 0 \tag{34}$$

The value of X may be obtained with the use of the computer program entitled ROOTS.

1. Consider the reaction

$$CO(g) + H_2O(g) \rightarrow CO_2(g) + H_2(g)$$

where $K_p = 10$ at $690°K$; $K_p = 4.4088$ at $900°K$; and $K_p = .3756$ at $1500°K$.

(a) Calculate the equilibrium partial pressures where the gases are maintained at $900°K$ and mixed with initial partial pressures as follows: CO, 1.0 atmosphere; H_2O, 1.0 atmosphere; CO_2, 1.0 atmosphere; and H_2, 1.0 atmosphere. [Use Equation (34).]

(b) Calculate the equilibrium partial pressures where the gases are maintained at $900°K$ and mixed with initial partial pressures as follows: CO, 2.0 atmospheres; H_2O, 1.0 atmosphere; CO_2, 0.2 atmosphere; and H_2, 0.1 atmosphere. [Use Equation (34).]

(c) Calculate the equilibrium partial pressures where the gases are maintained at $1500°K$ and mixed with initial partial pressures as follows: CO, 1.0 atmosphere; H_2O, 1.0 atmosphere; CO_2, 1.0 atmosphere; and H_2, 1.0 atmosphere. [Use Equation (34).]

(d) Calculate the equilibrium partial pressures where the gases are maintained at 1500°K and mixed with initial partial pressures as follows: CO, 2.0 atmospheres; H_2O, 1.0 atmosphere; CO_2, 0.2 atmosphere; and H_2, 0.1 atmosphere. [Use Equation (34).]

(e) Calculate the equilibrium partial pressures where gases are maintained at 690°K and mixed with initial partial pressures as follows: CO, 2.0 atmospheres; H_2O, 5.0 atmospheres; and <u>no</u> CO_2 or H_2. [Use Equation (34).]

(f) Calculate the equilibrium partial pressures where gases are maintained at 690°K and mixed with initial partial pressures as follows: CO_2, 2.0 atmospheres; H_2, 5.0 atmospheres; and <u>no</u> CO and H_2O. [Use Equation (34).]

2. Given the reaction

$$N_2(g) + 3H_2(g) \rightarrow 2NH_3(g)$$

where the $Kp = 1.64 \times 10^{-4}$ at 400°C.

If 20 atmospheres of N_2 is mixed with 30 atmospheres of H_2, what are the equilibrium pressures?

3. Given the chemical reaction

$$2H_2O(g) \rightarrow 2H_2(g) + O_2(g)$$

where the $Kp = 3.649 \times 10^{-23}$ at 1500°K.

(a) Calculate the equilibrium partial pressures for the above reaction when water vapor, initially at one atmosphere pressure, is allowed to decompose. What is the percentage of decomposition for the water vapor?

(b) Calculate the equilibrium partial pressures for the above reaction when water vapor, initially at 100 atmospheres pressure, is allowed to decompose. What percentage of the water vapor is decomposed at equilibrium?

11.12 TEMPERATURE DEPENDENCE OF Kp

An alternate method for calculating the equilibrium constant for a chemical reaction is based on the equation

$$\ln Kp = \int \frac{\Delta H_T^o}{RT^2} dT \qquad (35)$$

Classical Thermodynamics

When Equation (12) is substituted into the above equation, the integrated form of Equation (35) becomes

$$\ln K_p = \frac{\Delta a}{R} \ln T + \frac{\Delta b}{2R} T + \frac{\Delta c}{6R} T^2 - \frac{I}{RT} + I' \qquad (36)$$

where I' represents a new integration constant with a value of -5.27006 for the carbon monoxide and water vapor reaction. The antilog of log K_p then yields the value of K_p at any desired temperature.

1. Calculate the equilibrium constant for the reaction of carbon monoxide and water vapor at 300, 400, 500, $\cdots$ 1500°K. [Use Equation (36) and appropriate integration constants.]

2. Calculate the equilibrium constant for the following reaction at 400, 500, $\cdots$ 1500°K.

$$H_2O(g) \rightarrow H_2(g) + \tfrac{1}{2} O_2(g)$$

3. The formation reaction for ammonia gas has an enthalpy of formation of $-46,190$ joules per mole which remains essentially constant over the temperature range 300 to 1000°K. The equilibrium constant for the same reaction

$$N_2(g) + 3H_2(g) \rightarrow 2NH_3(g)$$

is 1.64×10^{-4} at 400°C. Use this information to approximate the equilibrium constant at 300, 400, $\cdots$ 1000°K. [Integrate Equation (35) where ΔH^o is considered to be constant and solve for the value of K_p.]

REFERENCES

1. The heat capacities in Table 11.1 are based on values found in Spencer, H.M. and Justice, J. L., Journal of the American Chemical Society, **56**, 2311 (1934); H. M. Spencer and G. N. Flannagan, Journal of the American Chemical Society, **64**, 2511 (1942).

2. The thermodynamic values found in Tables 11.2, 11.3, and 11.4 are from "Selected Values of Thermodynamic Properties", National Bureau of Standard Circular 500; U. S. Government Printing Office, Washington, D. C. (1952).

3. The heat capacity data for N_2 gas from 0 to 298.15°K is based on published results, Giauque, W. F. and Clayton, J. O., Journal of the American Chemical Society, **55**, 4875 (1933).

CHAPTER 12

Statistical Thermodynamics

12.1 TRANSLATIONAL PARTITION FUNCTION

Translational motions of gaseous molecules are quantized according to the particle-in-a-box model with energies expressed as

$$\varepsilon_x = \frac{n_x^2 h^2}{8ma_x^2} \qquad n_x = 1, 2, 3 \cdots \qquad (1)$$

$$\varepsilon_y = \frac{n_y^2 h^2}{8ma_y^2} \qquad n_y = 1, 2, 3 \cdots \qquad (2)$$

$$\varepsilon_z = \frac{n_z^2 h^2}{8ma_z^2} \qquad n_z = 1, 2, 3 \cdots \qquad (3)$$

where the energies are separated into three directional components. With each direction is associated a box length (a) and a quantum number (n). Planck's con-

Statistical Thermodynamics

stant is h and m is the mass of a single particle. The translational partition function in the x-direction (q_x) is given by

$$q_x = \Sigma \exp[-n_x^2 h^2 / 8ma_x^2 kT] \qquad (4)$$

where q_x is the one-dimensional partition function, k is the Boltzman constant, T is the absolute temperature, and where the summation occurs over all states. The summation is too lengthy for computer evaluation, therefore the summation is expressed in integral form as

$$q_x = \int_0^\infty \exp[-n_x^2 h^2 / 8ma_x^2 kT] \, dn_x \qquad (5)$$

The integral may be evaluated with the use of AREA. Equation (5) may also be integrated to give

$$q_x = \frac{(2\pi mkT)^{\frac{1}{2}} a_x}{h} \qquad (6)$$

Similar expressions can be derived for q_y, and q_z. The total translational partition function (q_{trans}) is, therefore

$$q_{trans} = q_x q_y q_z \qquad (7)$$

or

$$q_{trans} = \frac{(2\pi mkT)^{3/2} V}{h^3} \qquad (8)$$

where V is the volume of the container.

Consider again the partition function in the x-direction. The probability of finding the particle in a specific translational level is given by

$$P(n_x) = \frac{\exp[-n_x^2 h^2 / 8ma_x^2 kT]}{q_x} \qquad (9)$$

where q_x is defined by Equation (4), (5), or (6).

1. Calculate the translational partition function, q_x, for an HCl molecule confined to a box which is ten centimeters in length by using the program AREA and Equation (5)

    (a) at $20°K$.

    (b) at $300°K$.

    (c) at $2000°K$.

The following program statements will help to keep the exponents within the limits of the computer:

```
10 LET M = 6.053 E-26
20 LET H = 6.626 E-34
30 LET K = 1.381 E-23
40 LET T = 20
50 LET A = .01
60 LET B = H/M/K*H/(8*A↑2*T)
500 DEF FNF (X) = EXP(-B*X↑2)
```

The area under the curve becomes extremely small at large values of x, therefore large x-values must be avoided to prevent underflow problems.

2. Calculate the translational partition function, q_x, for an HCl molecule confined to a box which is ten centimeters in length by using Equation (6)

   (a) at $20°K$.

   (b) at $300°K$.

   (c) at $2000°K$.

   Compare calculated values of the partition function with results from Problem (1).

3. Calculate the probability of finding an HCl molecule at $300°K$ in each of the following states. Consider a one-dimensional, ten centimeter box.

   (a) $n_x = 10^0$ (g) $n_x = 10^6$

   (b) $n_x = 10^1$ (h) $n_x = 10^7$

   (c) $n_x = 10^2$ (i) $n_x = 10^8$

   (d) $n_x = 10^3$ (j) $n_x = 10^9$

   (e) $n_x = 10^4$ (k) $n_x = 10^{10}$

   (f) $n_x = 10^5$

   [Use Equation (9) and the value of q_x from Problem (1) or (2).]

4. PLOT the probability function given in Equation (9) for an HCl gas molecule

   (a) at $20°K$.

   (b) at $300°K$.

   (c) at $2000°K$.

Statistical Thermodynamics

12.2 ROTATIONAL PARTITION FUNCTION

Rotational motions of linear molecules are quantized according to the rigid rotator model, with energies expressed as

$$\varepsilon_{rot} = \frac{h^2}{8\pi^2 I} J(J+1) \qquad J = 0, 1, 2, 3, \cdots \qquad (10)$$

where J is the rotational quantum number and I is the moment of inertia of the molecule. The rotational partition function for a heteronuclear diatomic molecule is

$$q_{rot} = \sum_J (2J+1) \exp\left[-J(J+1)h^2/8\pi^2 IkT\right] \qquad (11)$$

where the summation is completed over all states and the degeneracy of each rotational level is $(2J+1)$. At sufficiently high temperatures, the summation in Equation (11) may be replaced by an integral so that

$$q_{rot} = \int_0^\infty (2J+1) \exp\left[-J(J+1)h^2/8\pi^2 IkT\right] dJ \qquad (12)$$

The integrated form of Equation (12) becomes

$$q_{rot} = \frac{8\pi^2 IkT}{h^2} \qquad (13)$$

The rotational partition function for a homonuclear diatomic molecule has the form

$$q_{rot} = \frac{8\pi^2 IkT}{2h^2}$$

The probability of finding a heteronuclear diatomic molecule in a specific state is given by

$$P(J) = \frac{(2J+1)\exp\left[-J(J+1)h^2/8\pi^2 IkT\right]}{q_{rot}} \qquad (14)$$

where q_{rot} is defined by Equation (11), (12), or (13). The average rotational energy of the molecule would then be given by

$$\varepsilon_{ave} = P(0)\varepsilon_0 + P(1)\varepsilon_1 + P(2)\varepsilon_2 + \cdots$$

where the ε's are defined by Equation (10) with subscripts identifying the quantum numbers and where the probabilities are defined by Equation (14).

1. Calculate the rotational partition function, q_{rot}, for an HCl molecule which has a moment of inertia of 27.0×10^{-48} kg m^2 by using Equation (11) and the sample program

    (a) at 20°K.

    (b) at 300°K.

    (c) at 2000°K.

```
10 LET I = 5.196 E-24
20 LET H = 6.626 E-34
30 LET K = 1.381 E-23
40 INPUT T
50 LET A = H/I/I*H/(8*3.1416↑2*K*T)
60 LET J = 0
70 LET S = 0
80 LET G = 2*J + 1
90 LET Q = G*EXP(-J*(J+1)*A)
100 LET S = S + Q
110 PRINT J,S
120 LET J = J + 1
130 GO TO 80
140 END
```

The I in the program is defined as the square root of the moment of inertia. The order of evaluation is important in statement 50 to avoid underflow problems. During each loop the value of J is incremented, the contribution to q_{rot} is calculated and added to an accumulator entitled S. On each cycle the current J and the value of S are printed. Watch the value of S on each cycle and when S reaches a maximum, terminate the program. This maximum value of S is the desired rotational partition function.

2. Calculate the rotational partition function, q_{rot}, for an HCl molecule which has a moment of inertia of 27.0×10^{-48} kg m^2 by using Equation (12) and the program AREA

    (a) at 300°K.

    (b) at 2000°K.

The following program statements will help to keep the exponents within the limits of the computer:

```
10 LET I = 5.196 E-24
20 LET H = 6.626 E-34
30 LET K = 1.381 E-23
40 INPUT T
50 A=H/I/I*H/(8*3.1416↑2*K*T)
500 DEF FNF(X)=(2*X+1)*EXP(-X*(X+1)*A)
```

Statistical Thermodynamics

The I in the program is defined as the square root of the moment of inertia. The area under the curve becomes extremely small at large values of x, therefore large x-values must be avoided to avoid underflow problems.

3. Calculate the rotational partition function, q_{rot}, for an HCl molecule which has a moment of inertia of 27.0×10^{-48} kg m^2. [Use Equation (13) and your own program.]

4. Calculate the probability of finding an HCl molecule in each of the occupied rotational states

   (a) at 20°K.

   (b) at 300°K.

   (c) at 2000°K.

   [Use Equation (14) and the values of q_{rot} as calculated in Problems (1), (2), and (3).]

5. PLOT the probability function described by Equation (14) over the range of occupied levels

   (a) at 300°K.

   (b) at 2000°K.

6. Calculate the average rotational energy of an HCl molecule at 300°K. Compare with kT.

12.3 VIBRATIONAL PARTITION FUNCTION

Vibrational motions of gaseous diatomic molecules are quantized according to the simple harmonic oscillator with energies expressed as

$$\varepsilon = h\nu(v + \tfrac{1}{2}) \qquad v = 0, 1, 2, 3, \cdots \qquad (15)$$

where h is Planck's constant, ν is a characteristic frequency and v is a quantum number. If the zero-point energy is subtracted from Equation (15) the temperature dependent vibrational energy is expressed as

$$\varepsilon_{vib} = h\nu v \qquad v = 0, 1, 2, 3, \cdots \qquad (16)$$

The vibrational partition function is then given by

$$q_{vib} = \sum_v \exp[-h\nu v/kT] \qquad (17)$$

where k and T are Boltzmann constant and absolute temperature respectively. The series in Equation (17) may also be expressed as

$$q_{vib} = \frac{1}{(1 - e^{-x})} \quad (18)$$

where

$$x = \frac{h\nu}{kT} \quad (19)$$

The probability of finding an HCl molecule in a specific state is given by

$$P(v) = \frac{\exp[-h\nu v/kT]}{q_{vib}} \quad (20)$$

Consequently, the average vibrational energy neglecting zero-point energy is

$$\varepsilon_{ave} = P(0)\,\varepsilon_0 + P(1)\,\varepsilon_1 + P(2)\,\varepsilon_2 + \cdots \quad (21)$$

where the ε's are defined by Equation (16) with subscripts identifying the quantum numbers and where the probabilities are defined by Equation (20). The same average vibrational energy is more frequently expressed in terms of the first Einstein function so that

$$\varepsilon_{vib} = kT\,\frac{x}{e^x - 1} \quad (22)$$

where x is defined by Equation (19).

1. Calculate the vibrational partition function, q_{vib}, for an HCl molecule which has a characteristic frequency of 8.568×10^{13} cycles per second. Use Equation (17) and the program which is listed below:

```
10 LET F = 8.568 E13
20 LET H = 6.626 E-34
30 LET K = 1.381 E-23
40 INPUT T
50 LET V = 0
60 LET S = 0
70 LET A = H*F/(K*T)
80 LET Q = EXP(-A*V)
90 LET S = S+Q
100 PRINT V,S
110 LET V = V+1
120 GO TO 80
130 END
```

Statistical Thermodynamics

The program increments the value of v to infinity. Terminate the program, however, when the value of S reaches a maximum. This is the value of q_{vib}. Determine the vibrational partition function

(a) at $20°K$.

(b) at $300°K$.

(c) at $2000°K$.

2. Calculate, using Equation (18), the vibrational partition function, q_{vib}, for an HCl molecule which has a characteristic frequency of 8.568×10^{13} cycles per second.

(a) at $20°K$.

(b) at $300°K$.

(c) at $2000°K$.

3. Calculate the probability of finding an HCl molecule in each of the occupied vibrational states

(a) at $20°K$.

(b) at $300°K$.

(c) at $2000°K$.

[Use Equation (20) and the values of q_{vib} as calculated in the above problems.]

4. PLOT the probability function described by Equation (20) over the range of occupied levels

(a) at $300°K$.

(b) at $2000°K$.

5. Calculate the average energy of an HCl molecule by using Equation (21) and compare to kT where

(a) the temperature is $300°K$.

(b) the temperature is $2000°K$.

Compare to results from Equation (22).

12.4 MOLAR ENTROPIES OF IDEAL GASES

The calculated entropies of an ideal gas contain contributions from translational, rotational, and vibrational motions. In each case the entropy contribution is calculated from

$$S = \frac{N\varepsilon}{T} + k \ln q^N \tag{23}$$

where N represents Avagadro's number. When Equation (23) is applied to translational motions, with a correction for the indistinguishability of gaseous molecules, the equation becomes

$$S_{trans} = \frac{3}{2} R + R \ln \left[(2\pi mkT)^{3/2} \frac{Ve}{h^3 N} \right] \tag{24}$$

where V is the molar volume of the gas

Equation (24) may be further simplified so that

$$S_{trans} = 12.471 \ln M + 20.785 \ln T - 9.686 \tag{25}$$

where the pressure is one standard atmosphere, M is the molecular weight of the gas in grams per mole, and S has units joules per mole.

When Equation (23) is applied to a mole of rotating diatomic gas molecules, the equation becomes

$$S_{rot} = R + R \ln \left[8\pi^2 IkT/\sigma h^2 \right] \tag{26}$$

where σ is a symmetry number with a value of 1 for heteronuclear diatomics and a value of 2 for homonuclear diatomics. The equation may be further simplified to

$$S_{rot} = -3.268 + 8.314 \ln I' + 8.314 \ln T - 8.314 \ln \sigma \tag{27}$$

where

$$I' = \frac{I}{10^{-46}}$$

and where S_{rot} is expressed in joules per mole per degree.

When Equation (23) is applied to the vibrations of a diatomic gas molecule, the equation becomes

$$S_{vib} = R \frac{x}{e^x - 1} + R \ln \frac{1}{1 - e^{-x}} \tag{28}$$

Statistical Thermodynamics 137

where the first term involves the familiar first Einstein function and the second term involves the third Einstein function.

The total entropy of a mole of ideal diatomic molecules is given by

$$S = S_{trans} + S_{rot} + S_{vib} \tag{29}$$

1. Calculate the translational, rotational, and vibrational contributions to entropy at 298.15°K and 1000°K for a mole of HCl gas. Compare the total entropy at 298.15°K to the standard entropy published in thermodynamic tables. The moment of inertia for HCl is 0.27×10^{-46} kg m^2. The fundamental frequency of HCl is 8.568×10^{13} cycles per second.

2. Calculate the translational, rotational, and vibrational contributions to entropy at 298.15°K and 1000°K for a mole of CO gas. Compare the total entropy at 298.15°K to the standard entropy published in thermodynamic tables. The moment of inertia for CO is 1.456×10^{-46} kg m^2. The fundamental frequency is 6.43×10^{13} cycles per second.

3. PLOT the value of the third Einstein function over a range of x from 0.01 to 10.

4. Calculate the third law entropy of a mole of Argon gas at 298.15°K and compare to standard tables.

5. Consider the reaction of gases

$$D + H_2 \rightarrow H + DH$$

where H and D are atoms of light and heavy hydrogen with only translational degrees of freedom. The moments of inertia for H_2 and DH are $.4603 \times 10^{-47}$ and $.6130 \times 10^{-47}$ kg m^2 respectively. The fundamental frequencies of H_2 and DH are 12.47×10^{13} and 10.87×10^{13} cycles per second, respectively. Calculate the $\Delta S°$ for this reaction at 298.15°K.

6. Consider the reaction of gases

$$H_2 + D_2 \rightarrow 2HD$$

The moments of inertia for H_2, D_2, and DH are $.4603 \times 10^{-47}$, $.9198 \times 10^{-47}$, and $.6130 \times 10^{-47}$ kg m^2 respectively. The fundamental frequencies of H_2, D_2, and HD are 12.47×10^{13}, 8.964×10^{13}, and 10.87×10^{13} cycles per second respectively. Calculate the $\Delta S°$ for this reaction at 298.15°K.

12.5 FREE ENERGIES AND EQUILIBRIUM CONSTANTS

Consider a general gas reaction such as

$$A + B \rightarrow C + D$$

The ΔG^o for such a reaction is given by

$$G^o = -RT \ln \frac{(q_C^o/N)(q_D^o/N)}{(q_A^o/N)(q_B^o/N)} \exp(-\Delta \varepsilon_0/kT) \quad (30)$$

where N is Avogadro's number, and the partition functions for each of the substances are defined at one atmosphere pressure as follows:

$$q_C^o = (q_{C,trans})(q_{C,rot})(q_{C,vib}) \quad (31)$$

$$q_D^o = (q_{D,trans})(q_{D,rot})(q_{D,vib}) \quad (32)$$

Similar expressions exist for q_A^o and q_B^o. The $\Delta \varepsilon_0$ is a term which accounts for the fact that the molecular states do not have a common zero. The equilibrium constant Kp relates to ΔG^o by the following expression:

$$\Delta G^o = -RT \ln K_p \quad (33)$$

The combination of Equations (30) and (33) produces an equilibrium constant directly in terms of the partition functions where

$$K_p = \frac{(q_C^o/N)(q_D^o/N)}{(q_A^o/N)(q_B^o/N)} \exp(-\Delta \varepsilon_0/kT) \quad (34)$$

The necessary partition functions are described in Sections 12.1, 12.2, and 12.3.

1. Calculate the equilibrium constant at $25^\circ C$ for the reaction

    $$D + H_2 \rightarrow H + DH$$

    where H and D are atoms of light and heavy hydrogen with only translational degrees of freedom. The moments of inertia for H_2 and DH are $.4603 \times 10^{-47}$ and $.6130 \times 10^{-47}$ kg m^2 respectively. The fundamental frequencies for H_2 and DH are 12.47×10^{13} and 10.87×10^{13} cycles per second respectively. The $\Delta \varepsilon_0$ in the calculation is -5.284×10^{-21} joules per molecule of H_2 reacted. Calculate also the ΔG^o for the reaction.

Statistical Thermodynamics 139

2. Calculate the equilibrium constant at $25°C$ for the reaction

$$H_2 + D_2 \rightarrow HD + HD$$

The moments of inertia for H_2, D_2, and HD are $.4603 \times 10^{-47}$, $.9198 \times 10^{-47}$, and $.6130 \times 10^{-47}$ kg m^2 respectively. The fundamental frequencies of H_2, D_2, and HD are 12.47×10^{13}, 8.964×10^{13}, and 10.87×10^{13} cycles per second respectively. The $\Delta\varepsilon_0$ in the calculation is 1.043×10^{-21} joules per molecule of H_2 reacted. Calculate also the $\Delta G°$ for the reaction.

3. Calculate the equilibrium constant at $25°C$ for the reaction

$$^{16}O_2 + {^{18}O_2} \rightarrow {^{16,18}O_2} + {^{16,18}O_2}$$

The $^{16}O_2$ has an internuclear distance of 1.207 angstroms, and a fundamental frequency of 4.665×10^{13} cycles per second. The other molecular species of oxygen have similar internuclear distances and similar force constants.

4. Combine the results of Section 4.5 with the results of Section 5.1 to calculate the ΔH for the reaction

$$D + H_2 \rightarrow H + D_2$$

5. Combine the results of Section 4.6 with the results of Section 5.2 to calculate the ΔH for the reaction

$$H_2 + D_2 \rightarrow 2HD$$

12.6 PERFECT CRYSTAL MODEL

The Einstein model for monatomic crystalline substances pictures the individual atoms as oscillators in each of three dimensions. The oscillations are similar to the motions of a vibrating diatomic molecule. The allowed energies of each oscillator in each of three directions are given by

$$\varepsilon_{os} = h\nu(n + \tfrac{1}{2}) \qquad n = 0, 1, 2, 3 \cdots \qquad (35)$$

where ε_{os} is the energy of each allowed level, h is Planck's constant, ν is a characteristic oscillating frequency and n is a quantum number. If the zero-point energies are subtracted from the energies described by Equation (35) the resulting energies are

$$\varepsilon = nh\nu \qquad n = 0, 1, 2, 3 \cdots \qquad (36)$$

The partition function for the oscillations in the x-direction is

$$q_x = \sum_n \exp[-nh\nu/kT] \qquad (37)$$

The same partition function may also be expressed as

$$q_x = \frac{1}{1 - e^{-\theta/T}} = \frac{1}{1 - e^{-x}} \qquad (38)$$

where

$$\theta = \frac{h\nu}{k} \qquad (39)$$

$$x = \frac{h\nu}{kT} = \frac{\theta}{T} \qquad (40)$$

The θ is called the Einstein characteristic temperature and the x is the variable which is used in evaluating Einstein functions. Expressions similar to Equations (37) and (38) can be derived for the partition functions of oscillations in the y and z directions. The complete partition function for a single oscillating atom is given by

$$q_{os} = q_x q_y q_z \qquad (41)$$

or

$$q_{os} = \frac{1}{(1 - e^{-x})^3} \qquad (42)$$

Molar thermodynamic functions for Einstein crystals are summarized below:

$$E_0 = 3/2\, Nh\nu \qquad (43)$$

$$(E - E_0) = 3RT \frac{x}{e^x - 1} \qquad (44)$$

$$C_v = 3R \frac{x^2 e^x}{(e^x - 1)^2} \qquad (45)$$

$$S = 3R \frac{x}{e^x - 1} + 3R \ln \frac{1}{1 - e^{-x}} \qquad (46)$$

Where E_0 is molar zero-point energy; $(E-E_0)$ is the temperature dependent oscillating energy and the quanity x is defined by Equation (40).

TABLE 12.1
Einstein Characteristic Temperatures For Several Solids Expressed in Degrees Kelvin

Substance	θ
Pb	65
Al	280
Diamond	1240

1. Calculate the entropies for lead, aluminum, and diamond at 298.15°K. Compare with thermodynamic tables.

2. Calculate the heat capacities for lead, aluminum, and diamond at 25, 50, 75, ··· 2000°K. Inspect the data and determine the approximate temperature where the heat capacity reaches the classical limit of 3R.

3. PLOT the heat capacities for lead, and diamond from a low temperature to approximately 500°K. Compare the plots.

CHAPTER 13

Kinetics

13.1 FIRST ORDER REACTIONS

Consider a general and simple reaction where the reaction is first order

$$A \rightarrow \text{Products}$$
$$C \qquad x$$

where C represents the ever decreasing concentration of A and x relates to the amount of product which is formed. Let C_0 represent the initial concentration of A. The differential equation may be written for the rate of loss for A, or for the rate of production for x

$$\frac{-dC}{dt} = kC \qquad (1a) \qquad \frac{dx}{dt} = k(C_0 - x) \qquad (1b)$$

where k is the first order rate constant. Both equations may be integrated where the integration constants are determined with the condition that $C = C_0$ or $x = 0$ when $t = 0$.

$$\ln C = -kt + \ln C_0 \qquad (2a) \qquad \ln(C_0 - x) = -kt + \ln C_0 \qquad (2b)$$

Both equations may be arranged as follows:

$$\frac{C}{C_0} = e^{-kt} \qquad (3a) \qquad \frac{C_0 - x}{C_0} = e^{-kt} \qquad (3b)$$

Kinetics

where the terms on the left represent the fraction of A which has reacted. Algebraic manipulations on the above equations result in

$$C = C_0 e^{-kt} \quad (4a) \qquad x = C_0(1 - e^{-kt}) \quad (4b)$$

Equation (4a) predicts the concentration of a reactant as a function of time. Equation (4b) follows the build-up of product as a function of time.

The chemist begins his kinetic studies by gathering concentration data at regular time intervals. One of his first concerns is the determination of the reaction order. His second concern is the determination of the reaction rate constant, k. If the kinetic data follow the above equations, then the reaction is said to be first order and the rate constant can easily be calculated using any of the above equations.

One method of verifying first-order reaction kinetics by computer becomes evident from rewriting Equations (2a) and (2b).

$$\frac{1}{t} \ln (C_0/C) = k \quad (5a) \qquad \frac{1}{t} \ln [C_0/(C_0 - x)] = k \quad (5b)$$

If the above equations are used to calculate k for each time interval, and if each calculation yields the same k within experimental error, then the reaction is classified as first order. If, however, the calculated k values change as the reaction proceeds, then the conclusion is that the reaction is not first order, and that further study is necessary to determine the order of the chemical reaction.

A second method of verifying first-order reaction kinetics by computer involves the direct use of Equations (2a) and (2b). Notice that both equations describe a straight line

$$y = mx + b \qquad (6)$$

When x is the abscissa, and y is the ordinate, the slope is m and the intercept is b. If the data points are plotted according to Equation (2a) or (2b) and the points fall onto a straight line within experimental error, then the reaction is classified as first order. If, on the other hand, the plotted points do not fit a straight line pattern, then further study is necessary to determine the reaction order. The computer program entitled LINEQ will fit a set of kinetic data points with a straight line, and it will determine the slope and intercept of the best line. It will be necessary to inspect the standard deviations, the correlation coefficient and the difference table before concluding that the data truly fit a straight line. If the data points are of a straight line nature, then the slope from the LINEQ program gives the rate constant for the first order reaction.

1. Consider the following reaction in the solvent, carbon tetrachloride, at $45°C$:

   $$N_2O_5 \rightarrow N_2O_4 + \tfrac{1}{2} O_2 \uparrow$$

   Typical kinetic data is presented below:

time (sec)	$C_{N_2O_5}$ (moles/liter)	$C_{N_2O_4}$ (moles/liter)
0	0.400	0.00
200	.353	.047
400	.311	.089
600	.274	.126
800	.242	.158
1000	.213	.187
1200	.188	.212
1400	.166	.234
1600	.146	.254
1800	.129	.271
2000	.113	.287

   The reaction is first order and has a rate constant of $6.29 \times 10^{-4} \text{ sec}^{-1}$.

   (a) Show that the reaction is first order by using Equation (5a).

   (b) Show that the reaction is first order by using Equation (5b).

   (c) Show that the reaction is first order by using Equation (2a) and LINEQ.

   (d) Show that the reaction is first order by using Equation (2b) and LINEQ.

2. The reaction

   $$N_2O_5 \rightarrow N_2O_4 + \tfrac{1}{2} O_2 \uparrow$$

   in carbon tetrachloride has a rate constant of $6.29 \times 10^{-4} \text{ sec}^{-1}$. The initial concentration of N_2O_5 is prepared to be 0.3 mole per liter.

   (a) PLOT the concentration of the N_2O_5 as a function of time using Equation (4a).

   (b) PLOT the concentration of N_2O_4 as a function of time using Equation (4b).

   (c) How many seconds are required for the concentration of the N_2O_4 to reach 0.15 mole per liter.

3. The reaction

   $$N_2O_5 \rightarrow N_2O_4 + \tfrac{1}{2} O_2 \uparrow$$

   in carbon tetrachloride has a rate constant of 6.29×10^{-4} at $45°C$. The initial concentration of the N_2O_5 is 0.5 moles per liter. How much time must elapse before the concentration of the N_2O_5 is diminished to 0.05 mole per liter.

Kinetics

4. Consider the following reaction at 593°K:

   $$SO_2Cl_2(g) \rightarrow SO_2(g) + Cl_2(g)$$

   The partial pressure of the SO_2 was equal to zero at $t = 0$. Kinetic data for the reaction is listed below:

time (min)	$P_{SO_2Cl_2}$ (atm)	P_{SO_2} (atm)
10	.7895	.0105
60	.7391	.0609
110	.6919	.1081
160	.6477	.1523
210	.6063	.1937
260	.5676	.2324
310	.5313	.2787
360	.4974	.3026
410	.4656	.3344

   (a) Verify that the reaction is first order.

   (b) Determine the rate constant.

   (c) Determine the partial pressures of the SO_2Cl_2 at $t = 0$.

   (d) Determine the time required for half of the SO_2Cl_2 to react.

5. Given data for the thermal polymerization of 1,3-butadiene,

   Butadiene $\rightarrow$ Dimer

t (min)	$P_{Butadiene}$ (mm Hg)
0	720.0
2	535.1
5	425.7
8	353.5
11	263.9
14	234.3
17	210.6
20	191.2
23	175.2
26	161.6

   Is this reaction first order?

13.2 SECOND ORDER IN ONE REACTANT

Consider a general reaction where the reaction is second order in one reactant

$$2A \rightarrow \text{Products}$$

Let C_0 be the initial concentration of A, and let C be the ever decreasing concentration of A. The differential equation for the loss of reactant A is given by

$$-\frac{dC}{dt} = kC^2 \qquad (7)$$

where k is the second order rate constant. The equation may be integrated and the integration constant determined at t = 0 so that

$$\frac{1}{C} = kt + \frac{1}{C_0} \tag{8}$$

The following rearrangements are useful when solving for C, k, and t respectively:

$$C = \frac{C_0}{C_0 kt + 1} \tag{9}$$

$$k = \frac{C_0 - C}{t C_0 C} \tag{10}$$

$$t = \frac{C_0 - C}{k C_0 C} \tag{11}$$

Equation (8) is the equation of a straight line. If kinetic data, which is plotted 1/C versus t, falls on a straight line, then the reaction is second order in one reactant only. The slope of the line is also the second order rate constant. Equation (10) may also be used to verify a second order reaction in one reactant only. If the values of k are calculated at a series of time intervals and each produces the same k within experimental error, then the reaction must be second order in one reactant.

1. Decomposition of acetaldehyde follows the equation

$$CH_3CHO(g) \rightarrow CH_4(g) + CO(g)$$

Kinetic data for the decomposition at $759°K$ is given below:

t(sec)	C_{CH_3CHO} (moles/liter)
0	.01500
100	.01370
200	.01198
300	.01064
400	.00957
500	.00870
600	.00797
700	.00735
800	.00683
900	.00637
1000	.00597

(a) Use Equation (8) and LINEQ to verify that the reaction is second order in acetaldehyde.

Kinetics

(b) Show that the reaction is second order by using Equation (10).

(c) Determine the rate constant for this reaction.

(d) Use PLOT to show the decrease of acetaldehyde concentration as a function of time.

(e) Determine the concentration of the acetaldehyde when t = 1500 sec.

2. Consider the following data for the dimerization reaction of butadiene at $599°K$:

t (min)	$P_{Butadiene}$ (mm Hg)
0	720.
2	535.1
5	425.7
8	353.5
11	302.2
14	263.9
17	234.3
20	191.2
23	175.2
26	161.6

Is the reaction second order with respect to butadiene?

13.3 SECOND ORDER—TWO REACTANTS

Consider a general reaction which is first order in each of two reactants

$$A + B \rightarrow \text{Products}$$
$$a_0-x \quad b_0-x$$

Let the initial concentration of A be a_0 and the initial concentration of B be b_0. The progress of the reaction is indicated by x which represents the amount of A (or B) which has reacted. The differential equation for the increase of x as a function of time is

$$\frac{dx}{dt} = k(a_0 - x)(b_0 - x) \qquad (12)$$

where k is the rate constant. The integrated rate equation becomes

$$\frac{1}{a_0 - b_0} \log \frac{(a_0 - x)}{(b_0 - x)} = kt + \frac{1}{a_0 - b_0} \log \frac{a_0}{b_0} \qquad (13)$$

with the integration constant evaluated at t = 0. The same equation may also be written as

$$\frac{1}{a_0 - b_0} \ln \frac{a}{b} = kt + \frac{1}{a_0 - b_0} \ln \frac{a_0}{b_0} \qquad (14)$$

where

$$a = a_0 - x$$
$$b = b_0 - x$$

A chemical reaction is first order in each of two reactants if the kinetic data fit a straight line as described by Equations (13) or (14). Equation (13) may also be rearranged to solve for k.

$$\frac{1}{t(a_0 - b_0)} \ln \frac{b_0(a_0 - x)}{a_0(b_0 - x)} = k \qquad (15)$$

If the value of k is calculated at each time during the kinetic experiment, and the values of k are the same within experimental error, then the reaction is first order in each of two reactants.

In the special case where initial concentrations a_0 and b_0 are equal, the equations developed in this section break down. The equations from Section 13.2 must then be used where $a_0 = b_0 = C$.

1. Consider the following reaction between propionaldehyde and hydrocyanic acid

$$C_2H_5CHO + HCN \rightarrow C_2H_5 - \underset{\underset{OH}{|}}{\overset{C \equiv N}{\underset{|}{C}}} - H$$

Given the following kinetic data at room temperature:

t(sec)	$C_{aldehyde}$ (moles/liter)	C_{HCN} (moles/liter)
0	0.100	0.060
79.02	.095	.055
170.3	.090	.050
276.9	.085	.045
403.4	.080	.040
556.0	.075	.035
744.4	.070	.030
983.8	.065	.025
1300.4	.060	.020
1744.4	.055	.015

(a) Show that the reaction is first order in aldehyde and first order in HCN by using Equation (14) and LINEQ.

(b) Verify that the reaction is first order in each of the two reactants by using Equation (15).

(c) Determine the $t_{\frac{1}{2}}$ for the reaction, the time required for the reaction to be half way toward completion.

Kinetics

2. Consider again the reaction of propionaldehyde and hydrocyanic acid with experimental kinetic data given as follows:

t	$C_{aldehyde} = C_{HCN}$
0 seconds	.1200 moles per liter
120	.1032
240	.0905
360	.0806
480	.0727
600	.0662
720	.0607
840	.0561
960	.0521
1080	.0487

(a) Verify that the reaction is described by the equation

$$\frac{1}{C} = kt + \frac{1}{C_0}$$

where C represents the aldehyde or the HCN concentration.

(b) Determine the rate constant for the reaction.

13.4 PSEUDO—ORDER RATE CONSTANTS

Consider a general reaction

$$A + B \rightarrow Product$$
$$a \quad b$$

where the concentrations of A and B are a and b respectively. The initial concentrations of the two reactants are a_0 and b_0. The general differential rate expression for this reaction is

$$-\frac{dA}{dt} = -\frac{dB}{dt} = ka^\alpha b^\beta \qquad (16)$$

where k is the rate constant; and where α is the reaction order for reactant A, and β is the reaction order for reactant B. A procedure will be described for calculating the reaction order, α, for A followed by the calculation of the reaction order, β, for B.

Kinetic data may be gathered for the reaction of A and B where the concentration of B is large compared to A. The concentration of B changes very little during the experiment; therefore, the value of b can be considered to be practically constant. Equation (16) may be written as

$$-\frac{dA}{dt} = (kb^\beta)a^\alpha = Ka^\alpha \qquad (17)$$

Where K is a pseudo order rate constant. By investigating the kinetic loss of A during the experiment by methods described in Sections 13.1, 13.2, and 13.3, the reaction order, α, and the pseudo rate constant K may be determined.

Next, kinetic data will be gathered for the reaction of A and B where the concentration of A will be large compared to B. The concentration of A changes very little during the experiment; however, the value of a can be considered to be practically constant. Equation (16) may now be written as

$$-\frac{dB}{dt} = (ka^\alpha)b^\beta = K'b^\beta \qquad (18)$$

Where K' is another pseudo order rate constant. By investigating the kinetic loss of B during the experiment by methods described in Sections 13.1, 13.2, and 13.3, the reaction order, α, and the pseudo rate constant K' may be determined. Thus the reaction orders for both reactants have been determined.

Finally the rate constant, k, in Equation (16) may be approximated from the definition of K and K' in Equations (17) and (18) respectively.

$$K = kb^\beta \quad \text{or} \quad k = \frac{K}{b^\beta} \qquad (19)$$

$$K' = ka^\alpha \quad \text{or} \quad k = \frac{K'}{a^\alpha} \qquad (20)$$

The following problems deal with the chemical reaction

$$C_2H_5CHO + HCN \rightarrow Products$$

1. The concentration of HCN is 0.8 moles per liter, which is large compared to the concentrations of the aldehyde. Determine the reaction order of the aldehyde and the pseudo rate constant as described in Equation (17).

t(sec)	$C_{aldehyde}$ (moles/liter)
0	0.0500
21	0.0414
33	0.0372
50	0.0320
61	0.0288
98	0.0210
112	0.0186
125	0.0166
140	0.0146
157	0.0126

Kinetics

2. The concentration of the aldehydes is 0.75 moles per liter, which is large compared to the concentration of the HCN. Determine the reaction order for the HCN and the pseudo rate constant as described in Equation (18).

t(sec)	C_{HCN} (moles/liter)
0	0.0400
22	0.0332
35	0.0298
57	0.0248
76	0.0212
83	0.0200
100	0.0174
123	0.0144
156	0.0110
178	0.0092

3. Combine the results of Problems (1) and (2) above and calculate the rate constant for the reaction. [Use Equations (19) and (20).]

13.5 DIFFERENTIAL METHOD FOR REACTION ORDERS

Consider a general reaction

$$A + B \rightarrow \text{Products}$$
$$a \quad b$$

where the concentrations of the reactants are a and b, and where the initial concentrations are a_0 and b_0. The general differential rate expression for this reaction is

$$-\frac{dA}{dt} = -\frac{dB}{dt} = ka^\alpha b^\beta \qquad (21)$$

where k is the rate constant and where the reaction orders are α and β for A and B respectively. In the differential method of determining reaction orders, the values of (dA/dt) or (dB/dt) must be determined experimentally. Recall that if the concentration of A is plotted versus times, then the slope of the curve is (dA/dt). Likewise, if the concentration of B is plotted against time, the slope of the curve is (dB/dt). These reaction velocities are most frequently determined during the very early portion of a kinetics experiment where t ≅ 0, because the plotted points define what is nearly a straight line.

To determine the reaction order of A in the above example, two experiments must be completed. The initial concentrations of B (b_0) will be identical in both experiments, whereas, the initial concentrations of A will be chosen to be different. The differential rate equations for both experiments can be written as

$$(-dA_1/dt)_0 = k(a_0)_1^\alpha b_0^\beta \qquad (22)$$

$$(dA_2/dt)_0 = k(a_0)_2^\alpha b_0^\beta \qquad (23)$$

where the subscripts refer to experiments 1 and 2. The subscript 0 indicates that the reaction velocities are being determined at $t \cong 0$. By dividing Equation (22) by Equation (23) and taking the logarithm we get

$$\alpha = \frac{\log[-dA_1/dt]_0 - \log[-dA_2/dt]_0}{\log[a_0]_1 - \log[a_0]_2} \qquad (24)$$

where the k and b_0 cancel out. All quantities in Equation (24) are experimentally determined, therefore the reaction order, α, for A can be calculated. A second set of two experiments where the initial concentrations of A (a_0) will be identical in both experiments and where the concentrations of B will be chosen to be different will yield the reaction order β. An expression similar to Equation (24) can be developed to calculate β

The following problems deal with the chemical reaction

$$C_2H_5CHO + HCN \rightarrow Products$$

1. Given data for two separate kinetic experiments where the concentration of the HCN is 0.08 mole per liter in both cases. Use the differential method as described in Equation (24) to solve for the reaction order in the aldehyde. Concentrations of the aldehyde are expressed in moles per liter. [Use LINEQ to determine reaction velocity.]

TRIAL I		TRIAL II	
time (sec)	$C_{aldehyde}$	time (sec)	$C_{aldehyde}$
0	.1200	0	.1000
11.2	.1188	14.6	.0987
22.7	.1176	30.8	.0973
32.5	.1166	41.5	.0964
41.5	.1157	50.0	.0957
47.6	.1151	63.6	.0946
59.0	.1140	75.0	.0937

2. Given data for two separate kinetic experiments where the concentration of the aldehydes is 0.075 mole per liter in both cases. Use the differential method and an equation similar to Equation (24) to solve for the reaction order in HCN.

Kinetics

Concentrations of the HCN are expressed in moles per liter. [Use LINEQ to determine reaction velocity.]

TRIAL III		TRIAL IV	
time (sec)	C_{HCN}	time (sec)	C_{HCN}
0	.1200	0	.1400
15.0	.1185	13.7	.1384
28.4	.1172	27.0	.1369
42.2	.1159	37.0	.1358
56.4	.1146	58.7	.1335
69.9	.1134	79.5	.1314
100.5	.1108	90.8	.1303

3. Combine the results of the two above problems to solve for the rate constant for the reaction.

13.6 ACTIVATION ENERGIES

The relationship between rate constants and temperature was initially described by Arrhenuis as follows:

$$k = A \exp[-E_a/RT] \quad (25)$$

where k is the rate constant, E_a is the activation energy, R is the gas constant, T is absolute temperature and A is a constant which is proportional to collision frequency. The same expression is often rewritten by taking the logarithm of both sides so that

$$\ln k = \frac{-E_a}{R} \frac{1}{T} + \ln A \quad (26)$$

Equation (26) is an equation of a straight line where ln k is plotted versus reciprocal absolute temperature. The activation energy is given by

$$E_a = -(\text{slope}) R \quad (27)$$

1. Consider the following reaction in carbon tetrachloride

$$N_2O_5 \rightarrow N_2O_4 + \tfrac{1}{2} O_2$$

Given the rate constants at several temperatures:

$t(°C)$	$k \times 10^4 \ (\text{sec}^{-1})$
20	.235
25	.469
30	.915
35	1.747
40	3.266
45	5.99

(a) Determine the values of A and B in the equation

$$\ln k = \frac{A}{T} + B$$

(b) Calculate the activation energy for the reaction.

(c) Calculate the rate constant for the same reaction at $15°C$.

2. The rate constant for the reaction of 1,3-butadiene to form a dimer can be described by the following equation:

$$\ln k = -12,400 \frac{1}{T} + 10.120$$

where k is expressed in $mm^{-1} \, min^{-1}$ and T is degrees absolute. The dimerization reaction is second order with respect to the butadiene concentration. Determine the k, and $t_{\frac{1}{2}}$ for the reaction at each of the following temperatures: 600, 610, 620, ... $700°K$. Assume that in each case of interest the reaction vessel initially contains a butadiene partial pressure of 720 mm of Hg.

13.7 AN INTERACTIVE KINETICS PROGRAM

The computer program RATES is designed to complete most kinetic calculations for reactions which follow the equations as listed below:

a. $\log C = -kt + \log C_0$

b. $\frac{1}{C} = kt + \frac{1}{C_0}$

c. $\frac{1}{a_0 - b_0} \log \frac{(a_0 - x)}{(b_0 - x)} = kt + \frac{1}{a_0 - b_0} \log \frac{a_0}{b_0}$

The equations describe reactions which are first order in one reactant only, second order in one reactant only, and first order in each of two reactants, respecitvely. In each case the k is the appropriate rate constant and t is the time. The variables C_0, a_0, and b_0 represent the reactant concentrations at $t = 0$; whereas the variables C, a, and b represent the reactant concentrations as the reaction progresses. The variable x represents the loss of reactant during the chemical reaction. The calculating options within the program are summarized below:

1ST ORDER : $LOG(C) = -K*T + LOG(CO)$

OPTION #	INPUT	OUTPUT	STOP LOOP
0	CO,K,T	C	TYPE T = 0.
1	CO,T,C	K	TYPE T = 0.
2	CO,K,C	T	TYPE C = 0.
3	K,T,C	CO	AUTOMATIC

Kinetics

2ND ORDER : $1/C = K*T+1/C0$

OPTION #	INPUT	OUTPUT	STOP LOOP
4	C0,K,T	C	TYPE T = 0.
5	C0,T,C	K	TYPE T = 0.
6	C0,K,C	T	TYPE C = 0.
7	K,T,C	C0	AUTOMATIC

1ST ORDER IN 2 REACTANTS: $LOG((A0-X)/(B0-X))/(A0-B0) = K*T+LOG(A0/B0)/(A0-B0)$

OPTION #	INPUT	OUTPUT	STOP LOOP
8	A0,B0,K,T	X,A,B	TYPE T = 0.
9	A0,B0,T,X	A,B,K	TYPE T = 0.
10	A0,B0,K,X	A,B,T	TYPE X = 0.

TYPE -1 FOR OPTION TO STOP PROGRAM ENTIRELY

The RATES program is designed to give assistance in the solving of many kinetic problems which are found in physical chemistry texts. It is further intended that the RATES program will be useful in the physical chemistry laboratory. The program is, however, a seasonal program and requires library storage for the short period of frequent use.

```
LIST
RATES

1    PRINT "INSTRUCTIONS (1=YES,0=NO)";
2    INPUT N
3    IF N=0 THEN 67
4    PRINT
5    PRINT "DICTIONARY OF VARIABLES (1=YES,0=NO)";
6    INPUT N
7    IF N=0 THEN 15
8    PRINT
9    PRINT "T = TIME","C = CONC.","K = RATE CONSTANT FOR REACTION"
10   PRINT "C0 = INITIAL CONC. FOR SINGLE SPECIES REACTION"
11   PRINT "A0 = INITIAL CONC. OF REACTANT A"
12   PRINT "B0 = INITIAL CONC. OF REACTANT B"
13   PRINT "X  = CONCENTRATION OF A0 AND B0 REACTED"
14   PRINT "A = A0 - X","B = B0 - X"
15   FOR I=0 TO 8 STEP 4
16   PRINT
17   GOSUB INT(I/4)+1 OF 76,78,80
18   GOSUB INT(I/4)+1 OF 23,23,30
19   PRINT
20   NEXT I
21   PRINT "TYPE -1 FOR OPTION TO STOP PROGRAM ENTIRELY"
22   GOTO 67
23   PRINT "OPTION #","INPUT","OUTPUT","STOP LOOP"
24   PRINT
25   PRINT I,"C0,K,T","C","TYPE T = 0."
26   PRINT I+1,"C0,T,C","K","TYPE T = 0."
27   PRINT I+2,"C0,K,C","T","TYPE C = 0."
```

```
28    PRINT I+3,"K,T,C","C0","AUTOMATIC"
29    RETURN
30    PRINT "OPTION #","INPUT","OUTPUT","STOP LOOP"
31    PRINT
32    PRINT 8,"A0,B0,K,T","X,A,B","TYPE T = 0."
33    PRINT 9,"A0,B0,T,X","A,B,K","TYPE T = 0."
34    PRINT 10,"A0,B0,K,X","A,B,T","TYPE X = 0."
35    RETURN
36    GOTO 67
37    PRINT "INITIAL CONC. OF A";
38    INPUT A0
39    PRINT "INITIAL CONC. OF B";
40    INPUT B0
41    IF ABS(A0-B0)>.000001 THEN 45
42    PRINT
43    PRINT "A0 = B0 : IT IS ADVISABLE TO USE ";
44    GOSUB 78
45    RETURN
46    PRINT "CONC. USED",X
47    A=A0-X
48    B=B0-X
49    PRINT "CONC. OF A",A
50    PRINT "CONC. OF B",B
51    RETURN
52    PRINT "CONC. USED",
53    INPUT X
54    RETURN
55    PRINT "INITIAL CONC.",
56    INPUT C0
57    RETURN
58    PRINT "PRESENT TIME",
59    INPUT T
60    RETURN
61    PRINT "PRESENT CONC.",
62    INPUT C
63    RETURN
64    PRINT "RATE CONSTANT",
65    INPUT K
66    RETURN
67    PRINT
68    PRINT "---------------"
69    PRINT
70    PRINT "OPTION";
71    INPUT L
72    PRINT
73    GOSUB L+1 OF 84,92,102,111,117,125,134,143,149,158,168
74    IF L#-1 THEN 67
75    STOP
76    PRINT "1ST ORDER : LOG(C) = -K*T+LOG(C0)"
77    GOTO 82
78    PRINT "2ND ORDER : 1/C = K*T+1/C0"
79    GOTO 82
80    PRINT "1ST ORDER IN 2 REACTANTS: LOG((A0-X)/(B0-X))/(A0-B0) ="
81    PRINT "  = K*T+LOG(A0/B0)/(A0-B0)"
82    PRINT
83    RETURN
84    GOSUB 76
```

```
85    GOSUB 55
86    GOSUB 64
87    PRINT
88    GOSUB 58
89    PRINT "PRESENT CONC.",C0*EXP(-K*T)
90    IF T#0 THEN 87
91    RETURN
92    GOSUB 76
93    GOSUB 55
94    PRINT
95    GOSUB 58
96    IF T=0 THEN 100
97    GOSUB 61
98    PRINT "RATE CONSTANT",(LOG(C0)-LOG(C))/T
99    GOTO 94
100   PRINT "PRESENT CONC.",C0
101   RETURN
102   GOSUB 76
103   GOSUB 55
104   GOSUB 64
105   PRINT
106   GOSUB 61
107   IF C=0 THEN 110
108   PRINT "PRESENT TIME",(LOG(C0)-LOG(C))/K
109   GOTO 105
110   RETURN
111   GOSUB 76
112   GOSUB 64
113   GOSUB 58
114   GOSUB 61
115   PRINT "INITIAL CONC.",EXP(K*T)*C
116   RETURN
117   GOSUB 78
118   GOSUB 55
119   GOSUB 64
120   PRINT
121   GOSUB 58
122   PRINT "PRESENT CONC.",1/(K*T+1/C0)
123   IF T#0 THEN 120
124   RETURN
125   GOSUB 78
126   GOSUB 55
127   PRINT
128   GOSUB 58
129   IF T=0 THEN 133
130   GOSUB 61
131   PRINT "RATE CONSTANT",(1/C-1/C0)/T
132   GOTO 127
133   RETURN
134   GOSUB 78
135   GOSUB 55
136   GOSUB 64
137   PRINT
138   GOSUB 61
139   IF C=0 THEN 142
140   PRINT "PRESENT TIME",(1/C-1/C0)/K
141   GOTO 137
```

```
142    RETURN
143    GOSUB 78
144    GOSUB 64
145    GOSUB 58
146    GOSUB 61
147    PRINT "INITIAL CONC.",1/(1/C-K*T)
148    RETURN
149    GOSUB 80
150    GOSUB 37
151    GOSUB 64
152    PRINT
153    GOSUB 58
154    X=A0*(1-EXP(K*T*(A0-B0)))/(1-A0/B0*EXP(K*T*(A0-B0)))
155    GOSUB 46
156    IF T#0 THEN 152
157    RETURN
158    GOSUB 80
159    GOSUB 37
160    PRINT
161    GOSUB 52
162    IF X=0 THEN 167
163    GOSUB 58
164    GOSUB 47
165    PRINT "RATE CONSTANT",LOG(B0*(A0-X)/A0/(B0-X))/(T*(A0-B0))
166    GOTO 160
167    RETURN
168    GOSUB 80
169    GOSUB 37
170    GOSUB 64
171    PRINT
172    GOSUB 52
173    GOSUB 47
174    PRINT "PRESENT TIME",LOG(B0*(A0-X)/A0/(B0-X))/(K*(A0-B0))
175    IF X#0 THEN 171
176    RETURN
177    REM
178    REM
179    END
```

REFERENCES

1. Data for the decomposition of nitrogen pentoxide was prepared from data found in Henry Eyring and Farrington Daniels, Journal of the American Chemical Society, 52, 1472 (1930).

2. Data for the dimerization of 1,3-Butadiene was prepared from data found in William E. Vaughan, Journal of the American Chemical Society, 54, 3863 (1932).

3. Data for the reaction of propionaldehyde and hydrocyanic acid was fabricated, based on data found in W. J. Svirbely and James F. Roter, Journal of the American Chemical Society, 75, 3106 (1953).

4. Data for the gas phase decomposition of acetaldehyde was prepared from data found in C. N. Hinshelwood and W. K. Hutchinson, Proceedings of the Royal Society, A111, 380 (1926).

CHAPTER **14**

Quantum Mechanics

14.1 THE PARTICLE IN A BOX

A first example of a quantum mechanical calculation frequently utilizes a particle in a one-dimensional box. The simplest of these problems defines the box to extend from $x = 0$ to $x = a$. The potential energy, $V(x)$, is defined to be zero within the box and defined to be infinitely large in the region outside the box. The time independent Schrödinger equation for the particle in the one-dimensional box has the form

$$\frac{-h^2}{8\pi^2 m} \left[\frac{d^2}{dx^2} \psi(x) \right] = E(x)\psi(x) \tag{1}$$

where h is Planck's constant, m is the mass of a single particle, and $E(x)$ is the energy of the particle. An acceptable solution to the differential equation has the form

$$\psi(x) = A \sin \frac{n\pi x}{a} \tag{2}$$

Quantum Mechanics

where n is a quantum number with integer values (n = 1, 2, 3, ···) and A is a normalization constant which relates to probability. The corresponding particle energies must, however, be quantized so that

$$E(x) = \frac{n^2 h^2}{8ma^2} \quad (3)$$

The probability of finding the particle within specified limits is defined by

$$P(x) = \int A^2 \sin^2 \frac{n\pi x}{a} \, dx \quad (4)$$

where P(x) is the probability function. The total probability is normalized by adjusting the value of A so that P(x) has a value of 1 or (100%) when the integral in Equation (4) is evaluated from x = 0 to x = a. Each allowed wave function for the particle in a box has a normalization constant defined by

$$A = (2/a)^{1/2} \quad (5)$$

The normalized wave function as described in Equation (2) may be used to calculate the average value or expectation value of a number of physical quantities. Certain expectation values for a particle in a one-dimensional box are defined below:

$$\bar{x} = \int (A \sin \frac{n\pi x}{a}) (x) (A \sin \frac{n\pi x}{a}) \, dx \quad (6)$$

$$\overline{x^2} = \int (A \sin \frac{n\pi x}{a}) (x^2) (A \sin \frac{n\pi x}{a}) \, dx \quad (7)$$

$$\overline{p^2} = \int (A \sin \frac{n\pi x}{a}) (\frac{-h^2}{4\pi^2}) \frac{d^2}{dx^2} (A \sin \frac{n\pi x}{a}) \, dx \quad (8)$$

where $\bar{x}$, $\overline{x^2}$, and $\overline{p^2}$ are expectation values for position, position squared, and momentum squared respectively. Equations (6), (7), and (8) are each integrated from x = 0 to x = a. The integration in Equation (8) cannot be completed until the wave function is differentiated twice. Each of these expectation values may be obtained by using the computer program entitled AREA.

Similar expressions develop when the particle in a three-dimensional box is considered. The wave function and quantized energies are listed for the case where the potential barrier is infinite any where outside the box as follows:

$$\psi(x,y,z) = (2/a_x)^{1/2} \sin \frac{n_x \pi x}{a_x} \cdot (2/a_y)^{1/2} \sin \frac{n_y \pi y}{a_y} \cdot (2/a_z)^{1/2} \sin \frac{n_z \pi z}{a_z} \quad (9)$$

$$E(x,y,z) = \frac{h^2}{8m} (n_x^2/a_x^2 + n_x^2/a_x^2 + n_x^2/a_x^2) \quad (10)$$

where n_x, n_y, and n_z are quantum numbers associated with the component x, y, and z directions and a_x, a_y, and a_z are the component lengths of the box.

1. Assume that an electron is placed into a one-dimensional box which is 10 nanometers in length. PLOT the wave function for each quantum state where n = 1, 2, 3, 4, and 5. Use the same number of plotted points in each case so that the plots can be more easily compared.

2. Assume that an electron is placed into a one-dimensional box which is 10 nanometers in length. PLOT the probability function for each quantum state where n = 1, 2, 3, 4, and 5. Use the same number of plotted points in each case so that the plots can be more easily compared.

3. Verify that the normalization constant for an electron in a one-dimensional and 10 nanometer box is in fact equal to $(2/a)^{\frac{1}{2}}$. [Use AREA and Equation (4).]

4. Assume that an electron is placed into a one-dimensional box which is 10 nanometers in length. Calculate the probability of finding the electron in the interval a = 4.5 to a = 5.5 nanometers in each quantum state where n = 1, 2, 3, 4, and 5. [Use AREA and Equation (4).]

5. Calculate the expectation value for the position of the electron in a 10 nanometer box when in the n = 1 quantum state. Repeat the same calculation when the electron is in the n = 2 quantum state.

6. Calculate the expectation value for x^2 of an electron in a 10 nanometer box when it is in the n = 1 quantum state. Do the same calculation when the electron is in the n = 2 quantum state.

7. Calculate the expectation value for p^2 when the electron is in a 10 nanometer box and in the n = 1 quantum state. Calculate the average kinetic energy by using the expression

$$\bar{E} = \frac{\bar{p^2}}{2m}$$

and compare to the result obtained from Equation (3).

8. Calculate the ten lowest possible quantized energy levels for a particle in a one-dimensional box where

   (a) An electron is confined to a 0.1 nanometer box.

   (b) A sixteen pound shot-put is confined to six inch shipping carton.

   (c) A nitrogen molecule confined to 10 meter room.

9. Calculate the ten lowest possible quantized energy levels for an electron confined to a three-dimensional box with

   (a) $a_x = a_y = a_z = 10$ nanometers.
   (b) $100a_x = 10a_y = a_z = 10$ nanometers.

10. Determine the probability of finding an electron in the center one-eighth of a 10 nanometer cube. [Evaluate the probability between these limits: $a_x/4$ to $3a_x/4$; $a_y/4$ to $3a_x/4$; and $a_z/4$ to $3a_y/4$.]

14.2 THE SIMPLE HARMONIC OSCILLATOR

The bond between the two atoms of a diatomic molecule can be looked upon as a flexible spring which follows Hooke's law,

$$f = kx \tag{11}$$

where f is the restoring force, x is the displacement from an average internuclear distance, and k is a proportionately constant which is frequently called the force constant. As the bond is stretched, the diatomic molecule builds up potential energy which is expressed as

$$V(x) = \frac{kx^2}{2} \tag{12}$$

where V(x) represents the potential energy. These classical results are frequently combined with the quantum mechanical result that the vibrational energies are quantized according to the expression

$$E = \frac{h}{2\pi} (k/\mu)^{\frac{1}{2}} (v + \tfrac{1}{2}) \tag{13}$$

where k is the force constant for the bond, μ is the reduced mass of a single molecule, and v is a quantum number with possible values of 0, 1, 1, 2, $\cdots$. When a bond is fully extended, or fully compressed, all the vibrational energy becomes potential energy. If the bond becomes fully extended or fully compressed the x in Equation (12) becomes x_{max} and the potential energy can be equated to the total vibrational energy in Equation (13). More explicitly,

$$x_{max} = \sqrt{\frac{h(v + \tfrac{1}{2})}{\pi (ku)^{\frac{1}{2}}}} \tag{14}$$

where x_{max} represents the classical amplitude of vibration of the diatomic bond.

The largest possible internuclear distance is the sum of the average internuclear distance and x_{max}. The shortest possible internuclear distance is the difference between the average internuclear distance and x_{max}. Closer inspection of Equation (14) shows that the diatomic molecule has a different amplitude of vibration in each quantum level.

The quantum mechanical investigation of the simple harmonic oscillator begins with the standard Schrodinger equation

$$\frac{-h^2}{8\pi^2 \mu} \left[\frac{d^2}{dx^2} \psi(x) \right] + \frac{kx^2}{2} \psi(x) = E\psi(x) \qquad (15)$$

where the potential energy is that of the simple harmonic oscillator. Equation (15) must be rather carefully rearranged so as to place it into a standard differential equation known as Hermites equation. When the energies are quantized according to the expression

$$E = h\nu(v + \tfrac{1}{2}) \qquad (16)$$

where ν is a fundamental frequency which is different for each diatomic molecule and v is a quantum number with values 0, 1, 2, 3, $\cdots$; the allowed wave functions are

$$\psi_v(q) = \sqrt{\frac{(2a/\pi)^{\tfrac{1}{2}}}{2^v \, v!}} \; H_v(q) \, \exp[-q^2/2] \qquad (17)$$

where q is a generalized coordinate defined as $q = (2a)^{\tfrac{1}{2}} x$ and $a = 2\pi^2 \nu\mu/h$. The x is displacement from an average internuclear distance, k is the force constant and μ is the reduced mass. The $H_v(q)$ represents an associated Hermite polynomial which is listed as follows:

$$H_{v=0} = 1 \qquad (18)$$

$$H_{v=1} = 2q \qquad (19)$$

$$H_{v=2} = 4q^2 - 2 \qquad (20)$$

$$H_{v=3} = 8q^3 - 12q \qquad (21)$$

$$H_{v=4} = 16q^4 - 48q^2 + 12 \qquad (22)$$

The following wavefunctions corresponding to v = 0, 1, 2, 3, and 4 are obtained by substituting the appropriate quantum numbers and the corresponding Hermite polynomial into Equation (17) and by changing to the variable x:

$$\psi_0(x) = (2a/\pi)^{1/4} \exp(-ax^2) \tag{23}$$

$$\psi_1(x) = (2a/\pi)^{1/4} 2a^{1/2} x \exp(-ax^2) \tag{24}$$

$$\psi_2(x) = (a/2\pi)^{1/4} (4ax^2 - 1) \exp(-ax^2) \tag{25}$$

$$\psi_3(x) = (8a^3/9\pi)^{1/4} (4ax^3 - 3x) \exp(-ax^2) \tag{26}$$

$$\psi_4(x) = (a/288\pi)^{1/4} (16a^2 x^4 - 24ax^2 + 3) \exp(-ax^2) \tag{27}$$

A corresponding probability function may be defined so that

$$P_v(x) = \int [\psi_v(x)]^2 \, dx \tag{28}$$

where the integral is evaluated on an interval of x ranging from a large negative value to a large positive value. Expectation values similar to Equations (6), (7), and (8) may also be defined. The expectation value for x, for example, is defined as

$$\bar{x} = \int \psi_v(x) \cdot x \cdot \psi_v(x) \, dx \tag{29}$$

The potential energy of vibrating diatomic molecules in high quantum states show substantial deviation from the SHO energy as described by Equation (12). An empirical equation by Morse provides a more realistic description for the potential energy at these high quantum states. The Morse potential energy equation has the form

$$V(x) = D_e [1 - \exp(-\beta x)]^2 \tag{30}$$

where x is the displacement from an average internuclear distance, D_e is the dissociation energy from the bottom of the potential well; and β is a constant which relates to molecular parameters. The quantized energy levels for a vibrating diatomic molecule as described by a Morse potential have the form

$$E = h\nu_c (v + \tfrac{1}{2}) - h\nu_c x_c (v + \tfrac{1}{2})^2 \tag{31}$$

where the ν_c represents a characteristic vibrational frequency and x_c is the anharmonicity constant.

Many of the following problems deal with the gaseous HCl^{35} molecule. This molecule has a force constant of 4.81×10^2 Newtons per meter. It has an average internuclear distance of 0.1275 nanometers. Another closely related molecule, the DCl^{35} molecule, can be considered to have approximately the same force constant and the same average internuclear distance. Other molecular properties are likely to be different because the DCl^{35} has a reduced mass of 3.16×10^{-27} kilograms per molecule as compared to the reduced HCl^{35} reduced mass of 1.63×10^{-27} kilograms per molecule. The fundamental frequency for HCl^{35} is 8.66×10^{13} cycles per second. The fundamental frequency for DCl^{35} is 6.21×10^{13} cycles per second.

1. PLOT the SHO potential energy function of gaseous HCl for values of x ranging from -0.05 nanometers to +0.05 nanometers. [Use Equation (12).] Superimpose onto the computer plot a pencil drawing of the allowed energy levels as described by Equation (13).

2. PLOT the SHO potential energy function of gaseous DCl for values of x ranging from -0.05 to +0.05 nanometers. [Use Equation (12).] Superimpose onto the computer plot a pencil drawing of the allowed energy levels as described by Equation (13).

3. Calculate the classical amplitudes of vibration for the HCl molecule in the quantum states v = 0, 1, 2, 3, 4, 5. [Use Equation (14).]

4. Calculate the classical amplitudes of vibration for the DCl molecule in the quantum states v = 0, 1, 2, 3, 4, 5. [Use Equation (14).]

5. PLOT the SHO wave function for the gaseous HCl molecule as described by Equation (23) to (27).

6. PLOT the square of the SHO wave functions for the gaseous HCl molecule as described in Equations (23) to (27).

7. The vibrating diatomic molecule as described by the quantum mechanical wavefunction has an amplitude of vibration which is somewhat larger than the value calculated with Equation (14). Calculate, therefore, for each quantum state of gaseous HCl the probability that the molecule will be within the classical limits of x. Use Equation (28), the program AREA, and the integration limits as defined below:

    (a) level v = 0; -0.01092 to +0.01092 nanometers.
    (b) level v = 1; -0.01891 to +0.01891 nanometers.
    (c) level v = 2; -0.02442 to +0.02442 nanometers.
    (d) level v = 3; -0.02889 to +0.02889 nanometers.
    (e) level v = 4; -0.03276 to +0.03276 nanometers.

Quantum Mechanics

8. Calculate the expectation value of x for the HCl molecule in the v = 2 quantum state. [Use Equation (29) and AREA.]

9. Calculate the expectation value of x^2 for the HCl molecule in the v = 1 quantum state. [Use an equation similar to Equation (29) and AREA.]

10. Calculate the expectation value of p^2 for the HCl molecule in the v = 0 quantum state. [Use an equation similar to (29) and AREA.]

11. PLOT the Morse potential function for gaseous HCl. The dissociation energy, D_e, has a value of 8.528×10^{-19} joules/molecule. The molecular constant, β, has a value of 1.74×10^{10} meters^{-1}. Plot the function from x = -0.042 to x = +0.458 nanometers. [Use Equation (30).]

12. Calculate the quantized energy levels for the HCl molecule by the SHO model. Calculate the quantized energy levels for the DCl molecule by the SHO model. [Use Equation (16).]

13. Calculate the quantized energy levels for the HCl molecule by the Morse potential model. The characteristic frequency is 8.96×10^{13} cycles per second; the anharmonicity constant is 0.0174.

14.3 THE HYDROGEN ATOM

The Schröedinger equation for the hydrogen atom has the form

$$\frac{h^2}{8\pi^2 m_1} \nabla_1^2 \psi_T + \frac{h^2}{8\pi^2 m_2} \nabla_2^2 \psi_T - \frac{e^2}{r} \psi_T = E\psi_T \tag{32}$$

where m_1 and m_2 are the masses of the electron and proton, respectively; and ∇_1 and ∇_2 are the differential operators. The solution of this differential equation requires that the wave function be separated into three component wave functions such that

$$\psi_T = \psi(\phi)\, \psi(\theta)\, \psi(r) \tag{33}$$

The quantum solution for $\psi(\phi)$ is possible at only certain values of the quantum number m. Listed below are normalized hydrogen functions for $\psi_m(\phi)$ where the m quantum numbers are 0, 1, -1, 2, -2 respectively:

$$\psi_0(\phi) = \frac{1}{\sqrt{2\pi}} \tag{34}$$

$$\psi_1(\phi) = \frac{1}{\sqrt{\pi}} \cos\phi \tag{35}$$

$$\psi_{-1}(\phi) = \frac{1}{\sqrt{\pi}} \sin \phi \tag{36}$$

$$\psi_{2}(\phi) = \frac{1}{\sqrt{\pi}} \cos 2\phi \tag{37}$$

$$\psi_{-2}(\phi) = \frac{1}{\sqrt{\pi}} \sin 2\phi \tag{38}$$

The value of m for an s orbital is 0; the values of m for p orbitals may be 0, 1, and -1; the values of m for d orbitals may be 0, 1, -1, 2, and -2. The ϕ angle ranges from $\phi = 0$ to $\phi = 2\pi$ and the functions are normalized so that

$$P_{m}(\phi) = \int_{0}^{2\pi} [\psi_{m}(\phi)]^{2} \, d\phi = 1 \tag{39}$$

The quantum solution of the hydrogen atom for $\psi(\theta)$ is possible at only certain quantum values of l and at only certain quantum values of m. Listed below are normalized hydrogen functions for $\psi_{l,m}(\theta)$ where l has possible values of 0, 1, and 2:

$$\psi_{0,0}(\theta) = \frac{1}{\sqrt{2}} \tag{40}$$

$$\psi_{1,0}(\theta) = \frac{\sqrt{6}}{2} \cos \theta \tag{41}$$

$$\psi_{1,\pm 1}(\theta) = \frac{\sqrt{3}}{2} \sin \theta \tag{42}$$

$$\psi_{2,0}(\theta) = \frac{\sqrt{10}}{4} (3 \cos^{2} \theta - 1) \tag{43}$$

$$\psi_{2,\pm 1}(\theta) = \frac{\sqrt{15}}{2} \sin \theta \cos \theta \tag{44}$$

$$\psi_{2,\pm 2}(\theta) = \frac{\sqrt{15}}{4} \sin^{2} \theta \tag{45}$$

The quantum number $l = 0$ represents the s orbit; the quantum number $l = 1$ represents the p orbit; and the quantum number $l = 2$ represents the d orbit. The θ angle varies from $\theta = 0$ to $\theta = \pi$ and the functions are normalized so that

$$P_{l,m}(\theta) = \int_{0}^{\pi} [\psi_{l,m}(\theta)]^{2} \sin \theta \, d\theta = 1 \tag{46}$$

The quantum solution of the hydrogen atom for $\psi(r)$ is possible at only certain quantum values of n and at only certain quantum values of l. Listed below are normalized hydrogen functions for $\psi_{n,1}(r)$ where n has possible quantum values 0, 1, and 2:

$$\psi_{1,0}(r) = 2a_0^{-3/2} \exp(-r/a_0) \tag{47}$$

$$\psi_{2,0}(r) = \frac{1}{\sqrt{8}} a_0^{-3/2} (2 - r/a_0) \exp(-r/2a_0) \tag{48}$$

$$\psi_{2,1}(r) = \frac{1}{\sqrt{24}} a_0^{-3/2} (r/a_0) \exp(-r/2a_0) \tag{49}$$

$$\psi_{3,0}(r) = \frac{2}{81\sqrt{3}} a_0^{-3/2} [27 - 18r/a_0 + 2(r/a_0)^2] \exp(-r/3a_0) \tag{50}$$

$$\psi_{3,1}(r) = \frac{4}{81\sqrt{6}} a_0^{-3/2} [6r/a_0 - (r/a_0)^2] \exp(-r/3a_0) \tag{51}$$

$$\psi_{3,2}(r) = \frac{4}{81\sqrt{30}} a_0^{-3/2} (r/a_0)^2 \exp(-r/3a_0) \tag{52}$$

where a_0 is a defined constant, the Bohr radius, and has a value of 0.0529 nanometers. The quantum number n = 1 represents the K shell; the quantum number n = 2 represents the L shell; and the quantum number n = 3 represents the M shell. The r distance is measured from r = 0 at the nucleus to r = ∞. The functions are normalized so that

$$P_{n,1}(r) = \int_0^\infty [\psi_{n,1}(r)]^2 r^2 dr = 1 \tag{53}$$

1. PLOT the $\psi(r)$ functions over the range of r = 0 to r = 1.0 nanometers. The radial wave functions can be summarized as follows:

    (a) Equation (47) is a description of the 1s orbit.
    (b) Equation (48) is a description of the 2s orbit.
    (c) Equation (49) is a description of the 2p orbit.
    (d) Equation (50) is a description of the 3s orbit.
    (e) Equation (51) is a description of the 3p orbit.
    (f) Equation (52) is a description of the 3d orbit.

2. PLOT the values of radial distribution curve for each of the radial wave functions over the range of r = 0 to r = 1.0 nanometers. The radial distribution function is defined as $[\psi(r) \cdot r]^2$.

3. Calculate the probability of finding the electron in a hydrogen atom somewhere between $r = 0$ and $r = 0.0529$ nanometers within each of the quantum states defined by Equations (47) to (52). [Use Equation (53).]

4. PLOT the $\psi(\phi)$ functions over the range $\phi = 0$ to $\phi = 2$. These wave functions can be applied as follows:

    (a) Equation (35) applies to any s, any p_z or any d_{z^2} orbital.

    (b) Equation (36) applies to any p_x or any d_{xz} orbital.

    (c) Equation (37) applies to any p_y or any d_{yz} orbital.

    (d) Equation (38) applies to any $d_{x^2-y^2}$ orbital.

    (e) Equation (39) applies to any d_{xy} orbital.

5. PLOT the value of $[\psi(\phi)]^2$ for each of the wave functions described by Equations (35) to (39).

6. Calculate the probability of finding the electron somewhere between $\phi = 3\pi/4$ and $\phi = 5\pi/4$ within each of the quantum states described by Equations (35) to (30). [Use Equation (39) to calculate the probabilities.]

7. PLOT the $\psi(\theta)$ functions over the range $\theta = 0$ to $\theta = \pi$.

    (a) Equation (40) describes any s orbital.

    (b) Equation (41) describes any p_z orbital.

    (c) Equation (42) describes any p_x or p_y orbital.

    (d) Equation (43) describes any d_{z^2} orbital.

    (e) Equation (44) describes any d_{xz} or d_{yz} orbital.

    (f) Equation (45) describes any $d_{x^2-y^2}$ or d_{xy} orbital.

8. PLOT the values of $[\psi(\theta)]^2 \cdot \sin\theta$ for each of the wave functions described by Equations (40) to (45).

9. Use the AREA program to verify that the wave functions which are described by Equations (40) to (45) are normalized. [Use Equation (46).]

Selected Answers To Exercises

1.13.2 The heat capacity for CO_2 gas at $500°K$ is 44.39 joules per mole.

1.13.3 The heat capacity for CO gas at $500°K$ is 30.28 joules per mole.

4.6.1 The heat capacity, in joules per mole, for CO_2 gas at $300°K$ is 40.16.

4.6.2 The function is best described by:
500 DEF FNF(X) = A + B*X + C*X↑(-2)
The heat capacity of CO gas at $300°K$ is 29.13 joules per mole.

4.6.3 The value of each function at x = 1 is listed below:
$f_1(x)$ = 0.5820
$f_2(x)$ = 0.9207
$f_3(x)$ = 0.4587

5.6.1 The heat required to increase the temperature of a mole of CO gas from 300 to $1000°K$ is 21,650 joules. The AREA program easily converges with 40 segments.

5.6.2	If the function is integrated from x = 0 to x = 10, then the probability equals 1.00. If the same function is integrated from x = 0 to x = 2.5, the probability is .0908 or 9.08 per cent.
5.6.3	The calculated values for the definite integrals are summarized as follows:

     (a) 5.953 with 50 segments on the interval.

     (b) 1.852 with 150 segments on the interval x = .000001 to x = 3.1416.

     (c) 0.5000 with 500 segments on the interval x = .000001 to x = 50.

     (d) 0.5000 with 1000 segments on the interval x = 1E-10 to x = 80.

6.6.1	In a TYPE 4 equation, the slope is also the value of A, and the intercept is B. Therefore A = -5,081 and B = 21.13.
6.6.2	In a TYPE 1 equation, b is the slope with a value of -0.03300; and RT is the intercept with a value of 24.45.
7.4.1	The coefficients which best describe the heat capacity of gaseous benzene from 400 to 2000°K are listed as follows:

     A = -3.138

     B = 9.392 x 10^{-2}

     C = -5.033 x 10^{-5}

     D = 9.758 x 10^{-9}

7.4.2	The coefficient A = 31.14; the coefficient B = -0.1110; and the coefficient C = 2.420 x 10^{-4}.
8.4.1	The four real roots of the polynomial are 1, -3, -4, and 9.
8.4.2	The equation $A + 2BP + 3CP^2 = 0$, has a root at P = 221.3 atmospheres, which represents a minimum in the compressibility plot. The equation $AP + BP^2 + CP^3 = 0$ has roots at P = 0 and P = 432 atmospheres. The gas is ideal at these two pressures.

Selected Answers To Exercises 173

8.4.3   The acid is 34.4 per cent dissociated.

9.1.1   The density for H_2 gas at $400°K$ and at 10 atmospheres pressure is approximately 0.614 grams per liter. The density of CH_4 gas at $400°K$ and at 10 atmospheres of pressure is 4.89 grams per liter.

9.2.1   Van der Waals pressure of a mole of CO_2 gas at 0.2 liters and at $273.15°K$ is 52.7 atmospheres. The ideal gas pressure for the same gas is 112.1 atmospheres.

9.2.2   The coefficients for CO_2 in Equation (11) are as follows: A, 200; B, -49.57; C, 3.593; D, -0.153378. Therefore, van der Waals volume of a mole of CO_2 gas at 200 atmospheres and at $500°K$ is 0.1682 liters. The ideal gas volume for the same gas is 0.205 liters.

9.3.1   The pressure by the virial equation for a mole of N_2 gas at 0.2 liters and $273.15°K$ is 111.1 atmospheres. The ideal gas pressure is 112.1 atmospheres.

9.3.2   Determine the virial coefficients for N_2 gas at $273.15°K$ as a test case of your program. Compare with the same coefficients listed in Chapter 9. The volume by the virial equation of a mole of N_2 gas at 200 atmospheres and $500°K$ is 0.2256 liters. The ideal gas volume for the same conditions is .205 liters.

9.3.3   The volume of a mole of N_2 gas at 10 atmospheres and at $273.15°K$ is 2.26 liters. The volume of a mole of N_2 gas at 10 atmospheres and at $500°K$ is 4.12 liters.

9.4.2   The Beattie-Bridgeman pressure for a mole of N_2 gas at 0.2 liters and $273.15°K$ is 110.6 atmospheres. The ideal gas pressure is 112.1 atmospheres. The virial equation in Section 9.3 yields a pressure of 111.1 atmospheres.

9.4.3   The Beattie-Bridgeman volume for a mole of N_2 gas at 200 atmospheres and $500°K$ is 0.2257 liters. The ideal gas volume is 0.205. The virial equation in Section 9.3 gives a volume of 0.2256 liters.

Selected Answers To Exercises

9.4.4 The pressure of a mole of N_2 gas at 273.15°K and at 10 liters of volume is 2.238 atmospheres.

9.5.2 Van der Waals constants for hydrogen gas are as follows: a, 0.1622; b, 0.02167; R, .06663. Excellent plots are obtained at 25°K where the plotting interval starts at 0.03 liters, and at 33.3°K where the plotting interval starts at 0.04 liters.

9.6.1 The PV data for a mole of CO_2 gas at 373.15°K can be described by a third degree polynomial where A = 31.11, B = -.1255, C = 3.727×10^{-4}, and D = -2.790×10^{-7}.

9.7.1 The Boltzmann plot for H_2 gas at 1500°K should include velocities up to 10,000 meters per second. The Boltzman plot for I_2 gas at 1500°K should include velocities up to 1000 meters per second.

9.7.2 No more than 200 segments are required over the integration interval. Approximately 20% of the I_2 molecules at 298.15°K have velocities between 0 and 100 meters per second.

9.7.3 The upper limit of integration is sufficiently large at 3000 meters per second. Two hundred segments on the integration interval are sufficient for convergence of the area. Increasing the temperature by 10°K does <u>not</u> produce twice as many activated molecules.

9.8.1 At one atmosphere pressure the collision properties for N_2 gas at 298.15°K are as follows: Mean free path is 6.53×10^{-8} meters, Z_1 is 7.20×10^9 collisions per molecule per second, and Z_{11} is 8.87×10^{34} collisions per second per cubic meter.

10.1.1 The v = 5 vibrational level for the Cl_2 molecule has an energy of 3058 cm^{-1} by the SHO model; the same v = 5 level has an energy of 2981 cm^{-1} by the anharmonic model.

10.1.2 The J = 5 rotational energy level for the I_2 molecule has an energy of 60.30 cm^{-1} by the rigid rotator model.

Selected Answers To Exercises

10.1.3 The transition energies from the anharmonic vibrational model are as follows: $v = 0$ to $v = 1$, 2886 cm^{-1}; $v = 1$ to $v = 2$, 2782 cm^{-1}; $v = 2$ to $v = 3$, 2678 cm^{-1}.

10.1.4 The transition energies from the centrifugal stretch rotational model are as follows: $J = 0$ to $J = 1$, 16.94 cm^{-1}; $J = 1$ to $J = 2$, 33.88 cm^{-1}; $J = 2$ to $J = 3$, 50.80 cm^{-1}.

10.1.5 Dissociation occurs at $v = 29$ for HCl; dissociation occurs at $v = 82$ for N_2.

10.2.1
```
10 INPUT B
20 INPUT T
30 LET NO = 1000
40 LET H = 6.626 E-27
50 LET K = 1.383 E-16
60 LET C = 3 E10
70 PRINT "J LEVEL", "RELATIVE NO."
80 FOR J=0 TO 100 STEP 1
90 LET W = B*J*(J + 1)
100 LET G = 2*J + 1
110 LET N = NO*G*EXP(W*H*C/(K*T))
120 PRINT J,N
130 NEXT J
140 END
```

10.2.2 If a distribution of rigid-rotating HCl molecules at 300°K has 1000 molecules in the $J = 0$ level, then there are 2400 molecules in the $J = 5$ level.

10.2.4 The average rotational energy for an HCl molecule at 300°K is very close to the classical average of kT, 4.143×10^{-14} ergs per molecule.

10.2.5 If a distribution of rotating HCl molecules has centrifugal stretch and has 1000 molecules in the $J = 0$ level, then there are 2406 molecules in the $J = 5$ level.

10.3.2 If the distribution of vibrating HCl molecules has 6.023×10^{23} molecules, then the $v = 0$ level contains 5.268×10^{23} molecules at 2000°K.

10.3.5 The average thermal vibrational energy of a single O_2 molecule at 300°K is 1.772×10^{-16} ergs. The value of kT at this same temperature is 4.143×10^{-14} ergs.

Selected Answers To Exercises

10.4.1 The first Einstein function has a value of 0.5820 at x = 1.0.

10.4.3 At $300°K$ the total thermal vibrational energy of a mole of O_2 gas is 10.73 joules, or 2.564 calories.

10.4.5 At $1615.22°K$, the thermal vibrational energy of a mole of N_2 gas is 4000 joules.

10.6.1 The heat capacity for a mole of HCl gas at $1000°K$ is 31.48 joules per mole per degree.

10.6.2 The heat capacity for a mole of Cl_2 gas at $1000°K$ is 36.98 joules per mole per degree.

10.6.3 The second Einstein function has a value of 0.5234 at 4000 reciprocal centimeters and at $2000°K$.

10.6.4 The second Einstein function has a value of 0.6695 at 1556 wave numbers and at $1000°K$.

11.1.2 The heat capacity for O_2 at $300°K$ is 29.21 joules per mole.

11.1.3 If the correct temperature is substituted into the polynomial as expressed in Table 11.1, the calculated heat capacity will be 34.55 joules.

11.2.1 The value of the second Einstein function at x = 1 is 0.9207.

11.2.3 The vibrational heat capacity of HCl gas at $500°K$ is 0.1415 joules per mole per degree.

11.2.4 The total heat capacity for CO_2 gas under a constant one atmosphere pressure is 44.61 joules per mole per degree at $500°K$.

11.2.5 The coefficients should be similar in magnitude to those found in Table 11.1.

11.4.1 The heat required to increase the temperature of a mole of CO_2 from $298.15°K$ to $900°K$ is 28,023 joules.

11.4.3 The cooling of a mole of CO gas from 900 to $298.15°K$ releases heat. Therefore ΔH for the cooling is -18,469 joules.

11.5.1 The heat of the reaction at $900°K$ is -35,830 joules.

11.5.2 The heat of the reaction at $1500°K$ is -29,870 joules.

Selected Answers To Exercises

11.5.5 Most handbooks of chemistry list the heat of vaporization of water at 373.15°K as 40,700 joules per mole.

11.5.6 The heat of the reaction at 900°K is -35,830 joules.

11.5.8 The heat of formation for CO_2 gas at 900°K is -394,350 joules.

11.5.9 The heat of formation for CO_2 gas at 1500°K is -395,370 joules.

11.6.1 The entropy change for solid phase II in changing from 35.61 to 63.14°K is 23.40 joules per mole per degree. Compare the entropy of N_2 at 298.15°K to the value from Table 11.3.

11.6.3 The third law entropy for a mole of CO gas at 900°K is 231.43 joules per mole per degree.

11.7.1 The entropy change for the reaction at 900°K is -33.20 joules per degree.

11.7.2 The entropy change for the reaction at 1500°K is -28.06 joules per degree.

11.8.2 The free energy change for the reaction at 900°K is -5,950 joules.

11.8.3 The free energy change for the reaction at 1500°K is 12,210 joules.

11.9.1 When delta P is 0.19 atmospheres (The partial pressure of CO is 0.81 atmospheres), the free energy of the reaction is -192 joules. If delta P is adjusted to be slightly larger, the free energy of the reaction becomes zero.

11.9.2 When delta P is -0.24 atmospheres (The partial pressure of CO is 1.24 atmospheres), the free energy of the reaction is 3.13 joules.

11.9.3 The equilibrium constant for the reaction at 900°K is 2.214.

11.9.4 The equilibrium constant for the reaction at 1500°K is 0.3756.

11.10.1 The equilibrium constant, Kp, has a value of 2.214 for this reaction at 900°K.

11.10.2 The decomposition of water at 1500°K is described by the following thermodynamic values: the enthalpy of the reaction is 250,570 joules; the free energy of the reaction is 164,230 joules; and the equilibrium constant is 1.909×10^{-6} (moles per liter)$^{1/2}$.

Selected Answers To Exercises

11.11.1 These problems are identical to those found in Section 11.9.

11.11.2 Expressions similar to those found in Equations (33) and (34) must be derived. When the correct partial pressures are found, the equilibrium constant will have a value of 1.64×10^{-4}.

11.11.3 The water vapor at 1 atmosphere pressure and $1500°K$ is approximately .02% dissociated.

11.12.1 The equilibrium constant for this reaction at $1000°K$ is 1.382.

11.12.2 The decomposition reaction for the water vapor has an equilibrium constant of 1.908×10^{-6} at $1500°K$.

12.1.1 Use these parameters in completing the numerical integration: $20°K$, limits of 0. and 10^{10}, 200 intervals; $300°K$, limits of 0. and 5×10^{10}, 200 intervals; $2000°K$, limits of 0. and 10^{11}, 200 intervals. The one-dimensional partition function for the translational motion of an HCl molecule at $20°K$ is 1.5468×10^{8}.

12.1.3 The probability of finding an HCl molecule at $300°K$ in the $n_x = 10$ level is 1.6692×10^{-9}. The probability for the $n_x = 10^{10}$ level is zero.

12.1.4 The probability of finding an HCl molecule in the $n_x = 5 \times 10^{8}$ level is: 1.764×10^{-12} at $20°K$; 9.659×10^{-10} at $300°K$; and 5.956×10^{-10} at $2000°K$.

12.2.1 The rotational partition function for an HCl molecule at $20°K$ is 1.733.

12.2.2 The value of the rotational partition function for HCl at $300°K$ by Equation (12) is 20.12. The value of the rotational partition function for HCl at $300°K$ by Equation (11) is 20.45. Equation (12) can only approximate the true partition function as defined by Equation (11).

12.2.3 The value of the rotational partition function for HCl at $2000°K$ by Equation (13) is 134.1. The value of the rotational function for HCl at $2000°K$ by Equation (11) is 134.4. Equation (13) only approximates the true value of the partition function.

Selected Answers To Exercises

12.2.4 The probability of finding an HCl molecule at $20°K$ in the $J = 1$ level is 39.0%.

12.2.5 The probability of finding an HCl molecule at $300°K$ in the $J = 1$ level is 13.3%.

12.2.6 The value of kT at $300°K$ is 4.143×10^{-21} joules per molecule. The average rotational energy is very close to the value of kT.

12.3.1 The vibrational partition function for HCl at $300°K$ has a value of 1.00000.

12.3.3 The probability of finding an HCl molecule at $2000°K$ in the $v = 1$ level is 11.2%.

12.3.5 The average energy of a single vibrating HCl molecule at $300°K$, neglecting the zero-point vibrations, is 6.427×10^{-26} joules. An identical result is obtained when Equation (22) is used.

12.4.1 The calculated entropy at $298.15°K$ is within several tenths of a joule of the published value.

12.4.3 The value of the third Einstein function at $x = 1$ is 0.4587.

12.4.5 The entropy change for the reaction at $298.15°K$ is 4.560 calories per degree.

12.5.1 The $\Delta G°$ for the reaction at $298.15°K$ is $-4{,}542$ joules.

12.5.2 The $\Delta G°$ for the reaction at $298.15°K$ is $-2{,}951$ joules.

12.5.3 The $\Delta G°$ for the reaction at $298.15°K$ is $-3{,}440$ joules.

12.6.2 Lead reaches a heat capacity of 24 joules per mole per degree at approximately $100°K$. Aluminum reaches the same heat capacity at approximately $420°K$.

13.1.1c The slope of an Equation Type 3 fit gives the negative value for the rate constant.

13.1.2 The concentration of N_2O_5 at $t = 2000$ seconds is 0.0853 moles per liter. The concentration of N_2O_4 at $t = 2000$ seconds is 0.2147 moles per liter.

13.1.4 The slope of an Equation Type 3 fit gives the negative value of the rate constant.

13.1.5 When the data points are inserted into Equation (5a), the calculated values of k are not constant. The reaction is not first order.

13.2.1 The slope of an Equation Type 5 fit gives the rate constant for the second order reaction.

13.2.2 The data points follow an Equation Type 5 fit in LINEQ. When data points are inserted into Equation (10), a constant value of k is obtained.

13.3.1 If Equation (14) is used, let
$$y = 1/(a_0 - b_0) \ln (a/b)$$
Calculate the value of y for each data point and use Equation Type 1 in LINEQ. The slope of the fit is also the second order rate constant. The rate constant is 0.0113 liters per mole per second.

13.3.2 Use a method from Section 13.2 to verify a second order reaction.

13.4.1 The reaction is first order in aldehyde. The pseudo rate constant is approximately $.0089 \text{ sec}^{-1}$.

13.4.3 The rate constant is approximately 0.0113 liters per mole per second.

13.5.1 The reaction velocity of Trial I is approximately -1.02×10^{-4} moles per liter per second.

13.5.3 The second order rate constant is 0.0113 liters per mole per second.

13.6.1 The slope of a Type 4 plot has a value of -12,080.

13.6.2 The $t_{1/2}$ for the reaction at $600°K$ is 52.8 minutes.

14.1.3 The integration of Equation (4) from x = 0 to x = 10E-09 gives an area of 1.00.

14.1.4 The probability of finding the n = 1 electron within the interval x = 4.5 to x = 5.5 nanometers is 19.8%. Approximately 20 per cent of the electron density is found within this interval.

14.1.5 The average position of the electron is at x = 5.000 nanometers in each of the quantum states.

Selected Answers To Exercises

14.1.6 The expectation value of x^2 for the electron in the n = 1 quantum level and in the 10 nanometer box is 2.8267×10^{-17} meters squared.

14.1.7 The average kinetic energy of the electron in the n = 1 quantum level and in the 10 nanometer box is 6.024×10^{-22} joules.

14.1.8 The lowest possible energy level corresponds to n = 1 in each case. The n = 1 energies are summarized as follows:

(a) 6.024×10^{-18} joules for the electron.

(b) 3.258×10^{-67} joules for the shot-put.

(c) 1.180×10^{-44} joules for the nitrogen molecule.

14.1.9 The first three energy levels within the cube are described by the following quantum numbers: (1,1,1), (1,1,2) or (1,2,1) or (2,1,1), and (1,2,2) or (2,1,2) or (2,2,1). The first three energy levels within the rectangular box are described by the following quantum numbers: (1,1,1), (1,1,2) and (1,1,3). The (1,1,1) quantum level in the rectangular box has an energy of 6.085×10^{-22} joules.

14.1.10 The probability of finding the n_x = 1 electron in the center half of a one-dimensional box in the x-direction is 0.818. The probability of finding the n_y = 1 electron in the center half of a one-dimensional box in the y-direction is 0.818. The probability of finding the n_y = 1 electron in the center half of a one-dimensional box in the z-direction is 0.818. The three-dimensional probability is therefore the product of these three probabilities or 0.547. The three dimensional probability for the center one-eighth is 0.125 for the (2,2,2) quantum state.

14.2.1 When the HCl bond displacement is +0.05 nanometers, the SHO potential energy is 6.013×10^{-19} joules.

14.2.2 When the DCl bond displacement is 0.05 nanometers, the SHO potential energy is 6.013×10^{-19} joules. The allowed vibrational levels are more closely spaced for the DCl as compared to the HCl molecule.

Selected Answers To Exercises

14.2.3 The maximum displacement for the HCl bond in the v = 0 quantum state is .0109 nanometers. The maximum internuclear distance is 0.1384 nanometers in the v = 0 state.

14.2.4 The maximum displacement for the DCl bond in the v = 0 quantum state is 0.0092 nanometers.

14.2.5 The vibrational wave functions have the following values at x = 0:

    (a) $\psi_0(x) = 227,360$

    (b) $\psi_1(x) = 0$

    (c) $\psi_2(x) = -160,770$

    (d) $\psi_3(x) = 0$

    (e) $\psi_4(x) = 139,230$

14.2.6 The vibrational wave functions squared have the following values at x = 0:

    (a) $\psi_0^2(x) = 5.169 \times 10^{10}$

    (b) $\psi_1^2(x) = 0$

    (c) $\psi_2^2(x) = 2.585 \times 10^{10}$

    (d) $\psi_3^2(x) = 0$

    (e) $\psi_4^2(x) = 1.938 \times 10^{10}$

14.2.7 The probability function of the v = 1 vibrational level evaluated between -0.0109 and +0.0109 nanometers is 0.843.

14.2.8 When the expectation value of x is determined between -10 to +10 nanometers with 50 segments on the integration interval, the expected value of zero is obtained.

14.2.9 The expectatin value of x^2 for HCl in the v = 1 state is 1.787×10^{-22} meters squared.

14.2.11 The Morse potential energy at x = -0.042 nanometers is equal to 9.887×10^{-19} joules.

Selected Answers To Exercises

14.3.1 The radial wave equations for the hydrogen have the following values at $r = 0.2$ nanometers:

(a) $\psi_{1,0}(r) = 1.1855 \times 10^{14}$

(b) $\psi_{2,0}(r) = -2.4711 \times 10^{14}$

(c) $\psi_{2,1}(r) = 3.0291 \times 10^{14}$

(d) $\psi_{3,0}(r) = -1.3098 \times 10^{14}$

(e) $\psi_{3,1}(r) = 1.2468 \times 10^{14}$

(f) $\psi_{3,2}(r) = 9.4987 \times 10^{14}$

14.3.2 The radial distribution functions for the hydrogen atom have the following values at $r = 0.2$ nanometers:

(a) $\psi_{1,0}^2(r) \cdot r^2 = 5.6217 \times 10^8$

(b) $\psi_{2,0}^2(r) \cdot r^2 = 2.4426 \times 10^9$

(c) $\psi_{2,1}^2(r) \cdot r^2 = 3.6702 \times 10^9$

(d) $\psi_{3,0}^2(r) \cdot r^2 = 6.8618 \times 10^8$

(e) $\psi_{3,1}^2(r) \cdot r^2 = 6.2178 \times 10^8$

(f) $\psi_{3,2}^2(r) \cdot r^2 = 3.6090 \times 10^8$

14.3.3 The probability of finding the 1s electron within a 0.0529 nanometer radius is 0.323.

14.3.4 The wave functions have the following values at $\phi = 1.5708$ radians:

(a) $\psi_0(\phi) = 0.39894$

(b) $\psi_1(\phi) = 0.00000$

(c) $\psi_{-1}(\phi) = 0.56419$

(d) $\psi_2(\phi) = -0.56419$

(e) $\psi_{-2}(\phi) = 0.00000$

14.3.6 The probability of finding an electron described by Equation (35) within the described limits is 0.409.

14.3.7 The wave functions have the following values at $\theta = 1.5708$ radians:

(a) $\psi_{0,0}(\theta) = 0.70711$

(b) $\psi_{1,0}(\theta) = 0.00000$

(c) $\psi_{1,1}(\theta) = 0.86603$

(d) $\psi_{2,0}(\theta) = -0.79060$

(e) $\psi_{2,1}(\theta) = 0.0000$

(f) $\psi_{2,2}(\theta) = 0.96825$

Index

INDEX OF COMPUTER TERMS

A

ABS function, 31
Addition, 3
AREA program, 39
   description, 41
   instruction for use, 43
   list, 44
Arithmetic operators, 3
Assignment operator, 3
Assignment statements, 2, 9
ATN function, 31

B

Backspace, 12
Break key, 12
BYE command, 13

C

CATALOG command, 15
Commands, 13
COMNT program, 39
Control C, 13

Core, active, 13
Correcting mistakes, 16
COS function, 31

D

DATA statement, 6
Data on paper tape, 12
Debugging, 25
DEF FNF statement, 32
Defining functions, 32
DELETE command, 14, 15
Deleting programs, 19
Descriptive comments, 39
Diagnostic messages, 24, 29
DIM statement, 54
Disk devices, 13
Division, 3

E

ELEVL program, 98
END statement, 2
E-Notation, 4
Error messages, 24, 27

Esc key, 12
EXP function, 31
Exponentiation, 3

F

FOR-NEXT statements, 7, 55
Functions, 30

G

GET command, 14
GFORRX program, 122
GO SUB statement, 55
GO TO statement, 5

H

HELLO command, 13
Hierarchy of operators, 8

I

ID code, 14
IF-THEN statement, 5
INPUT statement, 2

K

KEY command, 14
KILL command, 14

L

LET statement, 2
Library, 14
Linefeed key, 12
LINEQ program, 50
   description, 50
   instructions, 57
   list, 58
LIST command, 14
LOG function, 31

M

Magnetic tape, 13
Mathematical operators, 3
Multiplication, 3

N

NAME command, 14
NEXT statement, 7

O

OLD command, 14
Operators, 3

P

Parenthesis, 8
PASSWORD, 14
PLOT program, 30
   description, 32
   instruction, 33
   list, 34
POLEQ program, 65
   description, 65
   instructions, 66
   list, 67
PRINT statement, 2, 27, 40
Program, definition, 2
PUNCH command, 14
Punch-tape, 12, 19

R

RATES program, 154
READ statement, 6
REMARK statements, 40
RETURN statement, 12, 57
RND function, 31
ROOTS program, 74
   description, 74
   instructions, 76
   list, 77
RUN command, 13

S

SAVE command, 14
SCRATCH command, 14
SGN function, 31
SIN function, 31
SQR function, 31
Statements, 2
Statement numbers, 2
Subprograms, 55
SUBPRG program, 56
Subtraction, 3

Index 187

T

TAB function, 31
TAN function, 31
TAPE command, 14, 19
Teletype, 12
Test cases, 25
Time-sharing, 11
Tracing, 26
Trigonometric functions, 31

V

Variables:
    simple, 3
    subscripted, 53

W

Worded questions, 41

INDEX OF PHYSICAL CHEMISTRY TERMS

A

Activation energy, 93, 153
Amplitudes of vibration, 163
Anharmonicity constants, 96, 165
Atom, hydrogen, 167

B

Beattie-Bridgeman constants, 88
Boltzmann distribution function:
    rotational motions, 101
    translational motions, 92
    vibrational motions, 103
Boyle's law constant, 63, 91

C

Centrifugal distortion, 98
Characteristic wave numbers, 96
Collisions, molecular, 93
Compressibility, 1, 36, 70, 82, 91
Critical constants, 90
Critical state, 89
Crystal:
    energies, 140
    oscillations, 139
    perfect model, 139
    thermodynamics, 140

D

Degeneracy, 102
Diatomic molecules
    dissociation, 101
    rotations, 97
    vibrations, 95
Differential reaction orders, 151
Dissociation of bonds, 151
Dissociation of weak acid, 82

E

Einstein characteristic temperature, 140, 141
Einstein crystal model, 139
Einstein functions, 38
    first, 105, 140
    second, 107, 111, 140
    third, 137, 140
Electron wave functions, 167-170
Energy:
    activation, 93, 153
    average rotational, 103, 131
    average vibrational, 104, 134
    crystal, 140
    molecular, 95-108
    particle in a box, 128, 161
    potential, 160, 163
    potential, Morse, 165
    rotational levels, 131
    translational, 128
    vibrational levels, 104, 104, 133
    zero point, 104
Enthalpy:
    crystal, 140
    of formation, 114
    of heating, 45, 49
    of reaction, 113-116
    table of, 114
    statistical, 139
    of vaporization, 115
Entropy:
    crystal, 140
    equations for, 116, 118
    molecular contributions, 136
    of nitrogen, 117
    of reactions, 118
    standard, 117
    statistical, 136
    table of, 117
    third law, 116

Equations:
    Beattie-Bridgeman, 87
    Clausius-Clapeyron, 63
    Gibbs-Helmholtz, 120
    ideal gas, 83
    Boltzmann, 92, 101-104
    Schrödinger, 160, 164, 167
    Van der Waals, 23, 80, 84, 89
    virial, 85
Equations of state, gases, 63, 73, 83-89
Equilibrium constants, 124-127
    concentration, 125
    free energy and, 124
    partial pressures, 121, 125
    statistical, 138
    temperature dependence, 126
Expectation values, 161, 165

F

Force constants, 163
Free Energy:
    crystal, 140
    equations for, 119
    equilibrium and, 121
    formation, 119
    pressure and, 120
    standard, 119
    statistical, 138

G

Gases:
    collision frequencies, 93
    compressions of, 112
    expansions of, 112
    heat capacities, 38, 72, 111
    ideal, 83
    real, 84-89

H

Heat, 112
Heat capacities:
    crystals, 140
    experimental, 109
    gases, 38, 72, 107
    molecular contributions, 111
    nitrogen, 116
Heats of formation, 114
Heats of vaporization, 115, 117
Hermite polynomials, 164
Hooke's law, 163
Hydrogen atom, 167-170

I

Ideal gas, 83
Infrared spectra, 101

K

Kinetics, 142-158

L

Law, Boyle, 63, 91
Law, Hooke, 163
Least squares, 51, 65

M

Mean free path, 93
Mechanics:
    statistical, 128-141
    quantum, 160-170
Microwave spectra, 101
Molecular energies, 95-108
    rotational, 103, 131
    vibrational, 104, 134
    total, 106
    translational, 106
Molecular collisions, 93
Molecules, polyatomic, 106
Moments of inertia, 97
Momentum, 161
Morse potential, 165

N

Newton's method, 74
Normalization constants:
    hydrogen atom, 168-169
    particle in box, 161
    vibrating molecules, 165

P

Particle in box, 49, 128, 160
Partition functions:
    crystal, 140
    rotational, 131
    translational, 129
    vibrational, 133
Probability functions:
    hydrogen atom, 168
    particle in box, 161
    rotational, 131
    translational, 129
    vibrational, 135, 165

Index

Potential energy, 163, 165
Pseudo-order rate constants, 149

R

Radial wave functions, 169
Rate constants, 142–159
Reactions:
    first order, 142
    order, 142–159
    pseudo order, 149
    rate and temperature, 153
    second order, 145, 147
Reduced mass, 163
Relative populations:
    rotational levels, 102
    vibrational levels, 103
Roots of polynomial, 82
Rotations:
    average energy, 103, 131
    energy levels, 97
    spectra, 101
Rotational constants, 98
Rotor, rigid, 97

S

Schrödinger equation:
    hydrogen atom, 167
    particle in box, 160
    vibrating molecules, 164
Selection rules, 101
Spectra:
    infrared, 101
    microwave, 101
Speeds, molecular, 92
Statistical thermodynamics, 128–141
Symmetry number, 136

T

Temperature, Einstein, 140, 141
Thermodynamics:
    classical, 109–127
    statistical, 128–141

V

Van der Waals:
    constant, 85, 90
    equation of state, 23, 80, 84, 89
Vapor pressures, 63
Vibrations:
    amplitudes, 163
    anharmonic, 96, 165
    crystal, 139
    energies, 95
    harmonic, 95, 163
    selection rules, 101
Virial:
    coefficients, 86
    equation of state, 85
Volume, critical, 90

W

Wave functions:
    hydrogen atom, 167–169
    particle in box, 161
    vibrating molecule, 165
Wave mechanics, 160–170
Wave numbers:
    characteristic, 95
    fundamental, 106
Work, 112

Z

Zero-point energy, 104